Techniques *of*
Visual Persuasion

+ Create powerful images that motivate

New Riders | VOICES THAT MATTER™

Larry **JORDAN**

Techniques of Visual Persuasion: Create powerful images that motivate
Larry Jordan

Voices That Matter
www.voicesthatmatter.com
San Francisco, CA

Voices That Matter is an imprint of Pearson Education, Inc.
To report errors, please send a note to errata@peachpit.com

Alliance Editor: Laura Norman
Development Editor: Margaret S. Anderson
Technical Editor: Conrad Chavez
Senior Production Editor: Tracey Croom
Copy Editor: Kim Wimpsett
Compositor: Kim Scott, Bumpy Design
Proofreader: Becky Winter
Indexer: James Minkin
Cover Design: Chuti Prasertsith
Interior Design: Kim Scott, Bumpy Design

Section Opener Image Credits: Section 1: TessarTheTegu/Shutterstock, Section 2: Mykola Mazuryk/123RF, Section 3: hxdyl/123RF

ISBN-13: 978-0-13-676679-7
ISBN-10:   0-13-676679-X

1 2020

For my granddaughters, Sophie and Ellen Green,
two endlessly captivating distractions to writing.

And to Jane, for babysitting so I could write.

Acknowledgments

No book is written in isolation. I especially want to thank Laura Norman, executive editor for Pearson/Peachpit, for her continued support of this book; Margaret Anderson for her thoughtful editing; Conrad Chavez for his technical review and Kim Scott for the intricate jigsaw puzzle design of each page. This book has significantly benefited from their efforts.

I am deeply grateful to my intrepid team of beta readers: Jane and Paul Jordan, Annabelle Lau, Macaleigh Hendricks, and Noel Wong. They read and extensively commented on many (many) early drafts, helping me better define and explain these concepts. I'm deeply grateful for their thoughts. As well as my USC students over the years who helped me refine the most effective ways to teach these concepts.

To Stephen Roth and Paige Bravin, of Thalo LLC, for their encouragement and support of my training for many years.

I also want to thank our photos team: photographer, Janet Barnett, actors Kim Acunā, and Allison Williams, as well as Amy Camacho, for her skills at hair and makeup. Thanks to Janet, as well, for the use of her wine bar: Nabu. It is so exciting working with talented and enthusiastic people bringing the ideas in my imagination to life. You were amazing and fun to work with, as the images in this book attest.

Thinking of images, here's a quick shout-out to the talented photographers submitting images to Pexels.com. What a wealth of diverse and quality images! Many of the photos in this book were sourced from there.

Finally, a special thanks to Steve Mills (armpitstudios.com) and his image database: Image Chest. I used it throughout the writing of this book to track my images and the copyright/permission data that each required. His willingness to quickly fix bugs and add new features—which are now in the released product—was invaluable and deeply appreciated.

About the Author

Larry Jordan is a media producer, director, editor, author, and teacher. He's a member of both the Directors Guild of America and the Producers Guild of America, with network broadcast credits. In addition to a 20-year career in media, he's spent 30 years creating and marketing high-tech and training products from a variety of firms, including his own. He's written ten books on media and software.

His teaching background includes 12 years teaching visual storytelling and media software at USC in Los Angeles. He received the inaugural Industry Innovator Award presented by Digital Video in 2017.

Other Books by Larry Jordan

- *Dance for Television: A Production Handbook*
- *Editing Truths for Better Living*
- *Final Cut Pro HD: A Hands-on Guide*
- *Final Cut Pro 5: A Hands-on Guide*
- *Final Cut Pro Power Skills*
- *Final Cut Pro X: Making the Transition*
- *Edit Well: Final Cut Studio Techniques from the Pros*
- *Adobe CS Production Premium for Final Cut Studio Editors*
- *Adobe Premiere Pro Power Tips*

CONTENTS

PREFACE

So, where did you look first? At the photo or this text? Right. *Why* did you look at the photo first? Because I planned it that way. How did I plan it? *That* is what this book is about.

This book is about persuasion. More important, it is about persuading people remotely. How can we influence people when we are not in the same room with them? As you'll discover, persuasion is a one-on-one conversation between you and a member of your audience. In the past, we would do this face-to-face. Today, we need to do it remotely.

I started writing this book in early 2020, when all we were worried about in the United States was politics. I finished writing this book as the coronavirus altered everyday life, ending group activities and forcing most of us to live and work from home.

In the past, we could convince someone simply by meeting with them. Now, almost all our communication is remote. As the virus rages unchecked, devastating the world economy, visuals are more important than ever.

In today's world of "deep fakes," "fake news," and manufactured images, I thought a lot about whether to write this book. After all, showing how to create images to persuade pretty much defines much of what's wrong with today's society.

On the other hand, the folks who are causing trouble already know these techniques. My goals are different. To effect change—for good or ill—we need to persuade someone to make a decision to change. Persuasion is at the heart of all change. It is my hope that in sharing these techniques you will not only improve your own persuasive ability but better guard yourself from others who are using these techniques for nefarious purposes.

Knowledge is our best protection against manipulation. This is why I stress that telling the truth is a key element of any successful persuasive technique.

This book grew out of teaching the techniques of visual persuasion for more than a decade at the University of Southern California. My students helped me refine these ideas year after year. Now, I get to return the favor by creating a book with their notes already in it.

I also want to recommend the resources on my website: LarryJordan.com. Here you'll find thousands of tutorials and training videos that illustrate these subjects in-depth.

The role of persuasion isn't going away. Rather, it is morphing, moving online, and including more visuals. If you plan to persuade others, this book is essential reading. If you simply want to guard yourself against being persuaded, this book shows you the techniques others are using.

Never before in our society's history have the techniques presented in this book been more relevant. Stay safe and stay healthy.

Larry Jordan
Los Angeles, CA
April 2020

PERSUASION FUNDAMENTALS

SECTION GOALS

To communicate effectively today, we need to consider how a message *looks*, as well as what a message *says*. My goal is to get you thinking about images and design as part of your day-to-day communications, focusing on visuals as much as content. This doesn't mean you need to be an artist, because I certainly am not. Rather, I want to help you understand how the principles of design and media production can help you communicate more persuasively.

In this first section, we will cover fundamental concepts of persuasion and images. Once we cover the theory, we'll spend the rest of the book turning these ideas into action. The chapters in this section are:

- **Chapter 1: The Power of Persuasion.** What makes something persuasive? This also introduces two key concepts: defining your audience and the Call to Action.

- **Chapter 2: Persuasive Visuals.** This explains how images work, how audiences perceive images, and how we manipulate emotional content through camera placement and framing.

- **Chapter 3: Persuasive Writing.** This explains what a story is, how to write to support images, and what makes a word powerful.

- **Chapter 4: Persuasive Fonts.** Fonts evoke emotions. This illustrates how to use fonts that reinforce the story you are telling.

- **Chapter 5: Persuasive Colors.** Colors persuade subliminally. This shows what color is, the emotions it contains, and how to use color to enhance your visuals.

FOUR FUNDAMENTAL THEMES

- Persuasion is a choice we ask each individual viewer to make.

- To deliver our message, we must first attract and hold a viewer's attention.

- For greatest effect, a persuasive message must contain a cogent story, delivered with emotion, targeted at a specific audience, and end with a clear Call to Action.

- The *Six Priorities* and *the Rule of Thirds* provide guidelines we can use to effectively capture and retain the eye of the viewer.

THE SIX PRIORITIES

These determine where the eye looks first in an image:

1. Movement
2. Focus
3. Difference
4. Brighter
5. Bigger
6. In front

Persuasion, in the end, is about the other person making a decision....

Persuasion isn't (at least as we are talking about it) about power, coercion, or force.

It is about understanding, exploration, stimulation, and, ultimately, choice.

KEVIN EIKENBERRY[1]
LEADERSHIP EXPERT AND AUTHOR

CHAPTER 1

THE POWER OF PERSUASION

This chapter presents the fundamental concepts of persuasion, with a special focus on messages, audience, and the Call to Action. The goals of this chapter are to:

- Present an outline for this book
- Define "persuasion" and "visual persuasion"
- Explain the roles of message and emotion in persuasion
- Illustrate why we need to define and understand our audience
- Stress the importance of the Call to Action

[1] Eikenberry, K. (2014). Five Thoughts on Persuasion. https://blog.kevineikenberry.com/leadership-supervisory-skills/five-thoughts-on-persuasion/.

PERSUASION IS A CHOICE we present to a viewer to adopt our point of view. This point of view could be a product, an idea, social change, or something we want them to do on our behalf. Though we may be talking to large audiences, persuasion is always a one-on-one conversation between us and the viewer. In addition, persuasion today often occurs remotely. Rarely are we face-to-face with our audience.

Persuasion is a choice we present to a viewer to adopt our point of view.

Persuasion is not forced. We hold no power over our audience. We cannot demand. Instead, we need to convince others of the value of our ideas through the way we position, explain, and present them.

In the past, people wrote reports. Today, we share memes. Look at how you communicate: emojis in texts, images in email, and videos posted to Instagram, Facebook, or YouTube. We communicate using the shorthand of images.

Our society is awash in images because images are persuasive and entertaining. They don't require a lot of thought, the best ones communicate quickly, and we can share them with friends with a single click.

Persuasion is not just limited to consumers; it applies to business as well. The higher we rise in management, the more we rely on our communication skills. But, when persuading others, we face significant challenges in getting our message expressed, heard, and acted upon. There are a lot of barriers. Deadlines are short, planning and preparation are rarely sufficient, and, many times, we don't really know what we are doing.

As Jason Nazar, CEO of Comparably, writes, "Persuasion is the art of getting people to do things that are in their own best interest that also benefit you."[2]

If you need to convince someone about a new idea, a better way to work, or a technical problem—in fact, whenever you need to persuade anyone about anything—the techniques presented here can be invaluable.

Jeffrey Charles, CEO of Artisan Owl Media, writes, "I believe that influence is one of the top needs if we're going to be successful. ...You can't accomplish your goals without people. And if you're dealing with people, you need to be able to influence them."[3]

Successful persuasion requires us to control the story, the presentation, and the message to gain maximum impact with our target audience. This book is about learning how to control these aspects of a persuasive message using visual images.

[2] (Nazar, 2013)
[3] (Charles, 2016)

Dianna Booher has written a helpful book called *What More Can I Say?* on the process of persuasion. I found it useful in understanding how to best

(Image Credit: Free Vintage Illustrations of Gardens and Gardening, Free Vintage Illustrations)

explain the persuasive process. The word "persuasion" has connotations of Big Brother, mind control, and manipulation. But, as Booher writes, "*Persuading* is not a dirty word. It's not about manipulation. [Persuasion] is a neutral word. Whether it's good or bad depends on the [sender's] intellectual honesty, choice, purpose, and outcome."[4]

Persuasion is about choice. Yes, the viewer needs to make a choice on whether to accept our argument or not, but, more importantly, we need to make a choice in how we use these tools.

Professional filmmakers learned these techniques a long time ago. They apply them every day in creating commercials, feature films, and television programs. This is why what we watch today is so deeply moving. While the rest of us are not filmmakers, we can still leverage these techniques to meet the challenges of daily life. All too often, we post a message then wonder why no one is paying attention.

Persuasion is always a one-on-one conversation between us and the viewer.

How This Book Is Organized

My goal is to enable you to communicate more effectively using images. Throughout this book, I'll illustrate different, and ideally better, ways to create powerful, effective, and compelling images (and video) with the capability to persuade others.

[4] (Booher, 2015, p. xii)

Many times, the audience we need to convince is not in the same room with us. We need the best tools to persuade someone remotely. What's the most effective way to communicate with someone when we can't look them directly in the face? Right: images.

This book is divided into three major sections:

- "Persuasion Fundamentals"
- "Persuasive Still Images"
- "Persuasive Moving Images"

The first section—"Persuasion Fundamentals"—illustrates the core concepts that help create effective, persuasive images and video. (Most of these apply to both.) Learning these fundamentals will provide a firm foundation for the creative techniques we'll spend the rest of the book presenting.

The second section—"Persuasive Still Images"—looks at the process of creating effective still images, starting with presentations then expanding into photography and image editing.

The third section—"Persuasive Moving Images"—builds on everything we covered in the first two sections then adds tools to create compelling moving images. (And that movement adds a vast new layer of complexity.) This section includes audio, video, and motion graphics.

There's a lot to discover, but, at heart, we are simply trying to grab the attention of a viewer long enough to share our message and encourage them to act.

At heart, we are simply trying to grab the attention of a viewer long enough to share our message and encourage them to act.

Technically, "media," "film," "video," and "moving images" mean different things. However, most of us create moving images with our cell phones, not a $90,000 film camera. For this book, I'll use "media" to encompass both still and moving images, "film" and "video" will be used interchangeably, and "moving images" will differentiate between an image that is not moving (a "still") and an image that is (a "video").

Why Is Persuasion Necessary?

Most of us, when asked to name an activity that uses persuasion, would respond with "advertising" or "marketing." While true, this is a limited view of a broad activity. Persuasion applies to far more than just marketing. For example, it is used to get funding for a new idea, explain the benefits of a new law, or even explain to a 4-year-old why they need to take a bath.

In the past, we would generally effect change using face-to-face communication. In today's world, almost all persuasion happens remotely, where the person doing the persuading never meets the person they are trying to persuade. Because your audience has never met you, a key element of persuasion is built on trust. Your message is more likely to succeed if your audience trusts you. And trust cannot be built quickly.

Visual persuasion is the act of persuasion expressed using images, either still or moving. From broadcast and social media to posters taped on a wall, images surround us. As the speed of life continues to accelerate, images convey information faster and more powerfully than any other medium.

In fact, images today are more pervasive than the written word. It is increasingly clear that in today's visual and mobile society all of us need to know how to persuade. Visual communication and storytelling are now essential business and social skills because they cut through the noise to get our message heard. Persuasive images are no longer the exclusive province of the professional designer—they are a necessary communication skill for all of us.

We also need to recognize that an image is more than a picture; it is a visual representation of a message and an emotion. As you'll discover, effective images touch the mind and the heart. We need to move an audience. This requires us to, first, stop them long enough to actually see our message; second, deliver our message; and, third, take action. No presentation is sufficient simply providing information. It needs to include a core emotion, coupled with a compelling story, to drive our message deeply into the brain of the viewer.[5]

As the tools to create images continue to increase in power yet decrease in cost, more of us are creating images to persuade others. Sure, we can always hope to "go viral." But, wouldn't it be great if we could improve the odds? The problem is most of us don't know what we are doing. It's like handing a crayon to a toddler. They can make marks on paper, but to the rest of us, it's just a meaningless scribble.

This book grew out of a course I've taught for several years at the University of Southern California (USC), where students spend a semester learning a variety of media software. These aren't film students. Instead, they are computer programmers who want to learn better ways to communicate, engineering majors who want to learn design software, or business majors who are looking to expand their horizon beyond simply "running a business."

In school today, students are writing machines, creating theses, term papers, written reports of all descriptions. The problem is that today's society does not value long, written documents. Today's society values brevity, ideally with a twist at the end. Students are developing writing skills with limited commercial value after college.

Media today isn't a report, it's haiku. It isn't a paragraph, it's a bullet point. This is a hard transition for many of us to make. In almost all cases, we write too much. In my class, we focus on writing less and communicating more.

The pressure to communicate successfully is intense. The three broadcast television networks of 30 years ago have been replaced by an explosion of

Persuasive images are no longer the exclusive province of the professional designer—they are a necessary communication skill for all of us.

HAIKU A Japanese poem of seventeen syllables, in three lines of five, seven, and five syllables. Like all poetry, effective haiku requires using fewer words and making them work harder. We'll come back to this concept in Chapter 3, "Persuasive Writing."

[5] (Booher, 2015, pp. 132–133)

digitally delivered media. Streaming services are springing up like dandelions, sometimes welcome, sometimes not. The marketplace is flooded with content. Awards shows like the Oscars, Emmys, and Golden Globes, traditionally the domain of the major film studios and cable outlets, are seeing stiff competition from Netflix, Apple, Amazon, and other nontraditional distribution services.

With so much content to choose from, audiences are developing ever-shorter attention spans. Just a few years ago, we could comfortably watch a well-crafted 60-second commercial. Now, even a 15-second commercial seems too long. Producers struggle to connect with audiences and hold their attention even for short-duration ads.

Persuasion is an active conversation, built on trust, between you and your audience.

It's been said that there are only six or seven basic plots to choose from.[6] That means, as creators, we need to remember an audience always judges what they see today based upon what they've seen in the past. This past experience provides shortcuts we can use as communicators to trigger a feeling based only on the briefest of messages. For example, watching the first few seconds of a film trailer of a woman walking nervously down a dark street at night, the moon thinly veiled by clouds as a car screeches to a halt, pretty much tells you that we are not watching *Sesame Street*.[7]

You may think you are creating a poster for a school bake sale. But everyone will view and judge your poster based upon every other poster they've seen. To paraphrase my film editor friend, Norman Hollyn, everything we see is influenced by what we've seen before and what we see after.

This is a fundamental idea that runs throughout this book: Research has shown that if something looks good, people assume that it works "good." An image that looks "good" will be more effective than one that doesn't, even if the message they both contain is the same. This is especially true when you are trying to communicate with people who read less and watch more.

I remember reading that if the instructions for a task are well-designed and easy to read, the task is perceived as easier to do.[8] Even changing the font to make the text more readable (which we cover in Chapter 3) improves the perception that the task is simpler to complete.

What Is Persuasion?

[6] https://www.autocrit.com/blog/7-stories-world/
[7] (Ascher & Pincus, 2019, pp. 22–23)
[8] (Norman, 2004, p. 19)

Persuasion is a specific act designed to create change in an audience: a change in thought, behavior, laws, and so on. Persuasion is not violent. It is not a command. It is a choice you offer to someone else to follow your suggestions. In today's society, persuasion is pervasive.

To me, a persuasive message has five core components:

1	A clearly defined and articulated message
2	Containing a strong emotional hook
3	To a clearly defined audience
4	Ending with an explicitly stated Call to Action
5	All wrapped in a well-designed package

HOOK An element designed to catch people's attention. This could be any combination of movement, image, text, or sound.

Your goal is to create such a compelling desire in your audience that they want to do what you are suggesting. However, persuasion is as much what listeners *hear* as what you *say*. This is why understanding your audience is critically important.

Returning to Dianna Booher, "To influence someone to follow through on action, you have to start with the other person's reality—not yours."[9] This idea, that what *we* want is far less important than what the *viewer* wants, is one we'll come back to a little later in this chapter. For now, though, the reason we need to consider the needs of the viewer is that the best persuasion is not a monologue but a conversation.

"Pitching," Booher writes, "whether a formal sales pitch, an elevator pitch, or a crafted commercial—causes people to duck. A conversation, on the other hand, invites them to engage and exchange information. If you intend to persuade, make sure you're conversing, not pitching."[10]

Still, persuasion is meant to affect an audience. But, *how* do we do this? We don't change minds by shouting at people, as if the volume of our voice will somehow make a difference. Persuasion isn't shouting. It's connecting. It's listening. It's a two-way street.

As we will refer to a lot in this book, being persuaded is a choice the *viewer* makes. This means our efforts need to be focused first on attracting the attention of the viewer then delivering our message in such a fashion that they want to make the decision to change. And the most effective, most powerful, and, ultimately, most successful way to get people to pay attention is to appeal to their emotions as well as their intellect.

This idea of generating an emotional response in our audience is central to persuasion today. Donald Norman, in his book *Emotional Design*, writes, "Everything we do, everything we think is tinged with emotion, much of it subconscious. In turn, our emotions change the way we think, and serve as constant guides to appropriate behavior...."[11]

Persuasion isn't shouting. It's connecting. It's listening. It's a two-way street.

[9] (Booher, 2015, pp. 17, 19)
[10] (Booher, 2015, pp. 17, 19)
[11] (Norman, 2004, p. 7)

Since persuasion is a choice the viewer needs to make voluntarily, we need to find as many ways as possible to encourage them to make that choice. The best way to do so is to appeal to both their mind and their heart. This is why there is an ongoing focus in this book on the emotional content of words, images, fonts, and colors.

Our willingness to believe a message depends upon three factors: credibility, similarity, and power.

—ZOE ADAMS

Zoe Adams, PhD candidate at Queen Mary University in London, researched the intersection of consumer psychology and public health. What she discovered is that our willingness to believe a message depends upon three factors: credibility, similarity, and power.[14]

"Credibility," she writes, comprises trustworthiness and expertise and often refers to the believability of the source. "Similarity" indicates that attitude change was greatest for viewers who were told their thoughts were similar to other participants. And "power" refers to how the individual originating the message is perceived. High-power sources are perceived as more persuasive than less powerful sources, which is why testimonials are so commonly used today. We tend to respect the opinions of people who are perceived as well-known or experts.

Persuasion would be a lot easier, as well, if people were actually paying attention. But they aren't. Most of the time, they are distracted and thinking about something totally unrelated, like lunch. This means that, before they can be persuaded, they need to see our message, understand what we want them to do, then make the decision to do it.

One of the best ways to get someone's attention is with an image, which is why this book concentrates on techniques to make images more compelling. We need to stop people long enough for them to pay attention to our message.

What Is Visual Persuasion?

[12] (Norman, 2004, p. 60)
[13] (Booher, 2015, p. 115)
[14] (Adams, 2017)

I define "visual persuasion" as the process of convincing someone to take a specific action based primarily, though not exclusively, on an image or video. This is generally done remotely, where the person sending the message and the person receiving the message are not in immediate contact with each other.

Traditional examples of visual persuasion include television and magazine advertising, printed posters, and flyers. The opposite of visual persuasion is a white paper, where the message is delivered almost exclusively through text.

A good rule of thumb, when it comes to media, is "show, don't tell." Don't explain it with text, show it with an image. We will spend a lot of time in this book learning exactly how this is done, showcasing specific techniques you can put to use immediately.

Modern examples of visual persuasion still include advertising, of course, then add social media, business presentations, email, and texts. While we often think of email as personalized and text-based, think of all the emails you receive each day that are written for a mailing list and feature an image as the central focus of the message. Why? Because images are more persuasive than text.

Or, even more compelling, look at text messages today. They are filled with emojis, animojis, images, and memes of all sorts. What is an emoji if not a visualization of something we would normally write as text? We seem almost afraid to write—we would rather illustrate.

> *"Visual persuasion" is the process of convincing someone who is generally remote to take a specific action based primarily, though not exclusively, on an image or video.*

So, Did You See...?

Images are at the center of most of our conversations these days. For example, which do you tend to say more often: "So, did you read...?" or "So, did you see...?" Right. Our brains are hardwired to give priority to visual images.

For instance, which generates a faster and stronger emotional response, the statement "The tears rolled down the baby's cheeks" or **FIGURE 1.1**?

Of course, you looked at the baby first! In fact, you looked at the image before you read any text on this page. Any visual is far more powerful than any text.

While you may think of this as a "simple" photograph of a crying baby, the photographer is using specific techniques to deliver the maximum amount of emotional impact in this image. We examine these techniques in detail in Chapter 2, "Persuasive Visuals" and Chapter 7, "Persuasive Photos."

FIGURE 1.1 Of course you looked here first! Any visual is far stronger than any text. (Image Credit: pexels.com)

> *Each of us lives inside a personal bubble so dense that it takes a massive force to penetrate and grab our attention.*

In my definition of visual persuasion, I included the phrase that the sender and receiver were "not in personal contact with each other." This is important. If you are standing next to someone, there are stronger methods of persuasion than simply showing an image: a touch, the tenor of your voice, their perception of your emotions—all the cues we receive when we share the same space with someone else. However, those powerful cues are absent in digital media.

We are surrounded by technology today that "connects" us with others. But, while we've never been more connected, we have also never been more isolated. Each of us lives inside a personal bubble so dense that it takes a massive force to penetrate and grab our attention. In far too many cases, text isn't strong enough. A personal touch could break through, but too often, we live our lives through our screens, in emotional isolation from others. The only thing strong enough to penetrate our personal bubble is the visceral power of an image.

Reinforcing this lack of interest in text, one of my USC teaching assistants, Macaleigh Hendricks, told me: "In my experience, 99 percent of students are reading the absolute bare minimum of a chapter when it is assigned as homework. In fact, last semester, one classmate told me she would read the first and last sentences of every paragraph just to feel that she 'did the reading.'"[15]

Professional marketers know this, which is why so much of marketing today is image-driven.

Define Your Audience

The first assignment for the students in my USC class is to pick a message or theme that they want to explore during the semester. They are given a week to write a one-page description of their idea and turn it in. A key part of this first assignment is to define their audience. Most students reply, "Everyone!" And, why not? They want to attract as many people as possible.

But, if you were promoting healthy eating on *Sesame Street*, would you use the same language you would use to talk to your friends? Would you explain the music you like to your parents the same way you would explain it to your friends?

Messages work better when they are focused on specific audiences; different audiences require different messages. For example, imagine you are telling a story to a first-grader. Then, tell the same story to a high-school student. Finally, tell the same story in a business meeting. Even if the central message is the same, the story *must* change because the audiences are different.

[15] Macaleigh Hendricks, USC teaching assistant (Personal email)

This is a core point: stories (messages) do not exist in a vacuum. All stories are told with an audience in mind. You can't begin to tell a story until you first have a clear idea of who your audience is.

When I was younger, I studied to be a Dale Carnegie instructor. Dale Carnegie's best-selling book, *How to Win Friends and Influence People*, was released in 1936 and is still changing lives today. One of his central ideas is that "There is only one way under high heaven to get anybody to any-thing.... And that is by making the other person *want* to do it."[16]

Each of us needs the necessities of life—health, food, sleep, shelter, money—but after that, we need nourishment for our self-esteem. However, nourishing that self-esteem is different for different people. As communicators, we need to define who our audience is *and* define what they want.

During my Carnegie training, one of the most important lessons I learned was: WII-FM. ("What's in it for me?") We listen to everything through a filter that assesses how we can benefit from what we are being told.

Henry Ford wrote, "If there is any one secret of success, it lies in the ability to get the other person's point of view and see things from that person's angle as well as from your own."[17]

This is a critical point as we seek to persuade others. What we want *does not matter*! Of course, we want to improve ourselves; but who besides us cares? Nobody. Each member of our audience is too busy worrying about what matters to *them*.

This means we need to shift our thinking from what we need to what our audience wants. This audience-focus is critical to creating persuasively powerful, effective messages!

Harry A. Overstreet in his 1925 masterpiece of psychology, *Influencing Human Behavior*, wrote, "Action springs out of what we fundamentally desire...and the best piece of advice which can be given to would-be persuaders, whether in business, in the home, in the school, in politics, is: First, arouse in the other person an eager want.... The secret of all true persuasion is to induce the person to persuade himself."[18]

> *There is only one way...to get anybody to do anything.... And that is by making the other person want to do it.*
> —DALE CARNEGIE

"Why" Is a Key Question

Think about your reaction when someone unexpectedly asks you to buy something. How often do you stop what you are doing, carefully listen, then say, "Yes, please tell me more!" I suspect the answer is almost never. We are each far too busy to listen to unexpected sales pitches, so we tune them out.

[16] (Carnegie, 1961, p. 47)
[17] (Carnegie, 1961, p. 66)
[18] (Overstreet, 1925)

Nowhere is this more true than in avoiding web banners and online advertising; even our browsers are designed to block most advertising.

But, what if you are the one selling something—like an idea or improved process or, well, just about anything, actually? How can you get anyone to pay attention?

Well, the first thing is to have a compelling message, which is a combination of both text and images. That gets the viewer's attention. Next, though, is to give them a reason to act.

For any message to generate action, it needs to answer the question, "Why should I care?"—or—"Why is this important to me?" Remember the acronym from Dale Carnegie: WII-FM. "What's in it for me?" However, you don't have a lot of time—or words—to deliver that message. A 15-second video carries about 30 words. What can you say in 30 words that could possibly make a difference to the viewer?

As you'll learn throughout this book, the answer is to combine strong words with powerful images to cut through the clutter and drive your message home. We are not just "persuading." We are telling stories, guiding emotions, and setting a course of action for our viewers.

In the past, we could be subtle because people could connect the dots. Show them a picture of a dog then an image of dog food, and most people would say to themselves, "Oh! I maybe should buy this dog food for my pet."

Not today.

Today we are too distracted. So much so, that this spawned a new acronym: FOMO, the "fear of missing out." It's that anxious feeling you get when you haven't checked your phone in the last, say, five minutes. What are others doing? Are they having more fun than you? For my students, this need for reassurance is almost an addiction.

I see this in my classes. In mid-lecture, students are surreptitiously checking their cell phones, just to make sure they "stay in touch." Even when I ask them to put their phones away, the devices disappear for only a minute or two.

Another example is extreme multitasking, where an individual is doing several things at the same time. What I've discovered is that they may be "doing" multiple activities, but their brain is not deeply engaged with any of them. They are only paying superficial attention. This lack of engagement in any single activity means that they aren't truly paying attention to your message. The next notification will distract them.

We need to shift our thinking from what we need to what our audience wants.

The Call to Action

All this distraction brings me to the Call to Action (CTA), which is a term long used by marketers but is not well-known outside the ad business. The Call to Action is an explicitly stated behavior that you want the audience to do after seeing your image or watching your video.

The Call to Action forcefully tells the audience *what* you want them to do and *where* they need to go to do it. Calls to Action are not subtle. They are clear, direct, and delivered with all the force you can muster. Your image cuts into their "isolation bubble." The message tells them why they need to pay attention. The Call to Action tells them what they need to do next.

Remember, your audience is distracted. They don't want to do anything different. They need to be strongly motivated to change. Inertia rules.

It is impossible to overemphasize the importance of a clear and direct Call to Action. Even with a Call to Action, your audience may not pay attention. But, at least if you include it, they are more likely to do what you want.

Calls to Action should be used in as many types of marketing content as possible. Writing a good one is crucial and isn't that hard. Here are five steps, suggested by Ana Gotter, author and freelance business writer:

- **Focus on one goal.** The importance of this can't be overstated. Each ad campaign should focus on one primary goal.

- **Use action words.** Also known as verbs, action words are specific and motivating. "Shop," "Sign up," "Discover," "Try," "Watch," and "Start" are all examples. They directly tell customers what they should do next, which is what you want.

- **Pick the right formula for the right medium.** Tell your audience exactly what's in it for them. Sometimes, you'll want to keep your CTAs super short, but no matter what, make sure they're no more than five or six words. Anything longer takes away from the visual impact.

- **Decide if you want to go positive or negative.** This is an important part of the equation that plenty of people forget about. You can make your CTA a positive one or a negative one. Both are effective.

- **Prioritize brevity.** All of the best CTAs prioritize brevity. You don't have to follow a character count, but your CTA and surrounding text should never have any more words than are needed.[19]

I especially like her last point. We are not writing novels. We are creating haiku.

CALL TO ACTION An explicitly stated behavior that you want the audience to do after seeing your message. It tells them *what* you want them to do and *where* they need to go to do it.

Give your viewers every chance to successfully remember your message. When adding a URL as a Call to Action, consider capitalizing each word in the web address. Your computer doesn't care about case, while readers find "ThisIsATrulyGreatWebsite.com" much easier to read and remember than "thisisatrulygreatwebsite.com."

[19] (Gotter, 2019)

KEY POINTS

Here is what I want you to remember from this chapter:

- Persuasion is a choice the viewer makes, not a command you give.
- Persuasion is an active conversation, built on trust, between you and your audience.
- Persuasion is a one-on-one conversation, especially because today everyone is watching their own personal screen.
- Audiences are deeply distracted; focus your message on the audience you want to reach.
- Your message needs to explain why the audience should act the way you want them to act.
- A clear Call to Action is essential for your audience to know what you want them to do.

PRACTICE PERSUASION

Pretend you need to create a persuasive poster or video for something you like.

Use these to define your subject, audience, and message:

- What subject do you want to present?
- Who is your target audience, and why?
- Persuasion is about change. What change do you want your audience to undertake?
- Why should someone in your audience be interested in this subject; in other words, why should they care?
- How does your message change as your media changes?
- What is your Call to Action?

Write these into a one-page summary, no more, then put it somewhere you can find it. In a few chapters, we'll return to this for further discussion.

PEOPLE MAY ADMIRE YOUR WORK, BUT WILL THEY BUY?

Joe Torina
Producer/Director
Torina Media, Inc. (www.torinamedia.com)

In my experience, the effective use of text, narrative, and images is most immediately measured in direct response advertising (DR) and infomercials. Images are only part of the equation. All creative elements—sight, sound, script, etc.—exist for one purpose only: to connect with viewers on a visceral level. To sell! This is the secret sauce.

Home Shopping Network (HSN) in the mid-1980s was the silliest thing I had ever seen—hokey to the max. What I learned from my brief experience there, however, in terms of raw selling power, was phenomenal. At NBC, I was quite aware of ratings, total households, ADIs, and so on. It wasn't really until I stumbled upon retail television that I woke up to the business end of things.

HSN's business is the ability to sell, for example, $200 or $300 collectible dolls at the rate of $80,000 to $94,000 per minute! That was an eye-opener. And sales went on like that for three straight hours over both of HSN's two networks! This was the power of velocity selling, friends—the Amazon of its time.

I make infomercials for wealth building products, such as business opportunities, real estate education, and market trading software. These shows have to be extremely credible. They have to appeal to those who can afford to invest in their financial future. These are people who need to scrutinize, and they will accept nothing less than a solidly endorsed sales proposition.

DIRECT RESPONSE (DR) The promotional method in which a prospective customer is urged to respond immediately to the advertiser, through the use of a "device" (aka "Call to Action") which is provided in the advertisement.

For me, the Call to Action is the entire program itself. In my case, I motivate people to act days after they've seen the broadcast. This requires a certain potency in the messaging. A customer's commitment to travel to a seminar needs to keep fresh for a couple of days.

The shows work. Over a single five-year period and five different products, the infomercials and emulating direct mail generated $323.5 million in sales—an average of $64.7 million a year. Why did the shows perform this well? Because they had the look, feel, and texture of honesty and believability. My NBC background influenced the use of news magazine formats, which played well to our audience.

How, then, is believability and credibility achieved in a television sales proposition? It is basically in the honesty of production itself, in not fooling around with hype, distracting technique, and razzmatazz. It is in the virtue of using sound, pictures, music, dialogue, narration, b-roll, animation, and graphics straightforwardly, understandably, and intelligently—to tell an honest story in an honest way. Sounds quaint, but it's not.

In infomercials, the product is the star. Nothing gets in its way. Everything in the show supports it: talent, testimonials, music, animation, graphics, narrative…. We talk about it, we demonstrate it, and we mention its name countless times. I once counted how many times the product name was mentioned in one show. It was close to 100—about once every 17 seconds.

Testimonials for the product must be credible. That means no actors or rehearsed lines. Subjects must be completely real, everyday people. Again, it's more work producing this way, but it gets results.

Viewers sense dishonesty on the screen. Scripting an interview, for example, is a no-no in my book. It will never be believable, yet so many producers do it. It's extra work to log and cut, but anything I record, outside of voiceover narration, is strictly extemporaneous. Studio talent are guided only by bullet points. My testimonials, vignettes, studio segments, and vox pops (man-on-street interviews) are never scripted.

People find infomercials by accident. Wherever they may find your show, they must be compelled to stick with it. You must hold them for the duration!

There are a few basic infomercial ingredients:

- **Problem and solution.** This is a basic infomercial setup, especially at the intro and continued with variations throughout. "Are you struggling to save for retirement? How would you like to provide for your children's education and build for the future? Well, now you can!"

- **Features and benefits.** This is what's "written on the box." "Includes special training sessions to help you get up and running fast."

- **Demonstration.** This is key. Show how it works. "It's fun, it's fast, and it's easy!"

- **Overcome barriers.** It is imperative to destroy all possible objections in the viewer's mind, both conscious and subconscious. Express assumed barriers and quickly knock them down.

- **Call to Action.** Clearly and repeatedly tell the viewer what you want them to do.

As far as production is concerned, here are a few of my personal tenets:

- Pacing is key. Infomercials, more often than not, are joined midstream. Therefore, carry the pacing throughout. People will watch if things are moving along briskly.

- Sound is king—pictures follow. Good TV is good radio with pictures, they say. Keep the audio track moving; no dead air unless there's a real good reason for it.

- The product is exclusively the star. Always carry a notebook. Ideas come out of the blue when you least expect it. Lead with the action word, the verb. Time your script for maximum impact.

- Edit to relentlessly push the story ahead. Keep it moving, not frenetic of course, but moving—no lulls or confusing tangents.

- Voiceover may need extra lines in the edit. Be prepared.

- Music should never, ever be considered "wallpaper." That's a complete no-no. Music should always be treated as a predominant player in your story, its own character if you will.

- Interviews are best when kept conversational, not scripted.

The greatest compliment I ever got for one of my shows was from a program manager at an ABC affiliate. He refused to run the show in our scheduled fringe time because, as he said, "It looks too much like 'real' TV."

As filmmakers, we are highly conscious of elements the audience is only unconscious of.

NORMAN HOLLYN
FILM EDITOR
PROFESSOR, USC SCHOOL OF CINEMATIC ARTS

CHAPTER 2
PERSUASIVE VISUALS

CHAPTER GOALS

The emphasis of this chapter is to illustrate visual literacy. Specifically, to:

- Define "visual literacy" and other key visual composition terms
- Show how to catch and control the eye of the viewer
- Stress that the camera represents the eyes of the audience
- Illustrate the Six Priorities that determine where the eye looks first in an image based on composition
- Illustrate composition techniques we can use to make our images more compelling
- Illustrate the importance of the Rule of Thirds in framing
- Illustrate how camera position, blocking, framing, and angle affect the emotional response an image evokes in our audience
- Discuss whether sex sells

AN IMAGE DOESN'T EXIST IN A VACUUM. Understanding an image involves seeing the content, the environment, and the composition and placement of the image. In other words, every image is seen in the context of everything around it—and around us. The best images are not just "pictures"; they tell stories, like **FIGURE 2.1.**

FIGURE 2.1 Every image tells a story. Just like a story, the more focused the image, the more powerful story it tells. (Image Credit: pexels.com)

A good plan makes for good pictures.

Like any good story, an image improves when you think about:

- What you want to say
- Who you are saying it to
- What you want them to do after you say it

A good plan makes for good pictures.

You tell a funny story differently from a ghost story. A story told to children has different pacing than a story told to adults. Explaining a technical subject to people who are familiar with the subject is different from explaining it to people who are unfamiliar with it.

Not all of us are born storytellers, but all of us can improve with practice. Learning the "rules" of effective visual storytelling will also help. While "rules were made to be broken" remains true, still, to *break* the rules, you first need to know *what* those rules are.

What happens if you don't follow the rules? Well, most likely, no one would yell at you. No one would get hurt. No laws would be broken. Still, your message may lack impact and may not register with your audience. Given all the time and effort it took to create your image in the first place, can you really afford to overlook anything that might improve your results?

My core idea is that, as creators of visual messages, we are in control of where the viewer's eye looks within an image, the information the image delivers, and the emotion it evokes. By using these rules, we magnify the impact our visuals have on an audience. In this chapter, I'll illustrate the key visual concepts that we'll spend the rest of the book putting into practice.

Visual Literacy

Whether you are shooting photographs, designing motion graphics, or creating presentation slides, the image you create is what the audience sees. In other words, the image represents the point of view of the audience.

While this seems self-evident, it is hard to remember that when you change the position, the framing, or the content in front of the camera, what you are *actually* doing is changing what the audience sees and their perception of your message. When you move the camera, you are dragging the audience right along with you.

The quote from Norman Hollyn that started this chapter is one of my favorites. Norman was a highly talented film editor, as well as a professor at the USC School for Cinematic Arts. He and I spent four years coproducing and hosting the 32-part web series *2 Reel Guys* (www.2reelguys.com), which we described as "film school for filmmakers who didn't go to film school." Working with Norman as we wrote the scripts, then hosted each episode, was a deep-dive into the principles of filmmaking. My strengths are in media and technology; his were in storytelling and editing. What I learned from working with Norman was that the best visual media—even still images—tells a story. As communicators, our job is to figure out what that story is and the best way to tell it.

As an audience, we respond to these visual techniques without understanding *why* we respond the way we do. Very little in this chapter will be new to a filmmaker or graphic designer—they've been using these techniques for decades. However, what's important is that the rest of us *don't* know these techniques.

Let's set the scene by defining four key terms:

- Visual literacy
- Controlling the viewer's eye
- Composition
- Frame

As creators of visual messages, we are in control of where the viewer's eye looks within an image, the information it delivers, and the emotions it evokes.

Visual literacy. The term "visual literacy" was first coined in 1969 by John Debes, who founded the International Visual Literacy Association. The basic definition is: "The ability to read, write, and create visual images. It is a concept that relates to art and design, but it also has much wider applications. Visual literacy is about language, communication, and interaction. Visual media is a linguistic tool with which we communicate, exchange ideas, and navigate our complex world."[1]

Unlike real life, each still image and video is viewed through a frame, which creates the illusion of a boundary to every image.

Controlling the viewer's eye. Because each image tells a story, as communicators we need to make sure that the story the audience perceives is the story we are telling. Norman called this "manipulating" the viewer. I call it "guiding." Either way, we need to control the eye and mind of the viewer as much as we can to make our message as powerful as possible. Never cede control of your message to the viewer. Over the course of this book, you'll learn a wide variety of techniques to control your message.

Composition. This is the process of designing the content and technical aspects of an image. This could be as simple as taking a quick snapshot or as complex as shooting a scene for a major motion picture. Composition determines the choice and arrangement of the elements that go into creating an image. (Sometimes it is more important to decide what to leave *out* than what to put *in*.)

Frame. In real life, when we look around, we see everything all at once: left/right/up/down, close-up, and far away. But, when we look at any digital image, whether a video or still, we see it through a frame, an arbitrary boundary that determines the edges of the image.

What's truly important about the frame is that our brain assumes the edges of the frame are the limit of what is actually "there." The brain jumps to the conclusion that the frame represents the entirety of an image. Even though that's clearly not true, this false conclusion is impossible to resist. An actor, standing on the set, can easily see things outside the frame. For example, the camera, the crew behind the camera, the set and the lights above the set, the discarded props and costumes, and even actors—such as that blood-thirsty monster—that haven't appeared on set yet. But to our eyes, looking at the image, all we see is what's in the frame.

Because of the frame, the brain imposes artificial limits on the image. Those brain-imposed limits allow us, as communicators, to manipulate images to control the content, emotion, composition, and, ultimately, the story that each image tells. In reality, the heroine is not being attacked by the evil monster, but it is impossible to tell our brains that. The frame is a powerful instrument of persuasion.

[1] (Harrison, 2017)

The Six Priorities Determine Where the Eye Looks First

All things being equal, when looking at an image, we look at faces first. But, things are rarely equal, either by accident or by intent. What if there is more than one face in the image? What if there's only one face but it's obscured? What if the face is really small? That's where these *priorities* come in; they help us guide the viewer's eye to where we want it to look first.

I'm indebted to Norman Hollyn who first introduced this concept of visual hierarchy to me, though I've modified his order and definitions. Our brain is hardwired to process images in a specific order. We will refer to this list often as we create images because it determines the order in which our audience looks at elements within the image. (In fact, for easy reference, you'll find these listed at the start of each section in this book.)

FIGURE 2.2 Why does your eye see the black puzzle piece first? (Image Credit: pexels.com)

When looking at still or moving images, such as **FIGURE 2.2**, the eye goes through a "checklist" of where to look first, then second, and so on. I call this checklist "The Six Compositional Priorities that Determine Where the Eye Looks First." That is, ah, a mouthful, so I'll abbreviate this as the "Six Priorities." The eye looks at an image and the elements in it in a specific order, based on these priorities:

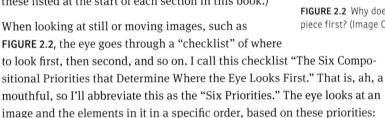

1	Movement
2	Focus
3	Difference
4	Brighter
5	Bigger
6	In front

If something is moving, our eyes look there first. If nothing moves, we look at that which is in focus. If everything is in focus, we look at that which is different. And so on. We look at elements that are higher on this list *before* we look at elements that are lower. These priorities are not only used for images or video; they also apply to how we look at the world around us in real life.

The Six Priorities guide the eye of the viewer so they see what you want them to see in the order you want them to see it.

Let me illustrate each of these.

It is impossible to overemphasize the importance of movement. Long before we learned to write, we were hunters—or being hunted. Our brains are hard-wired to pay attention to anything that moves. Our first thought is, "Is *that* food, or are *we* food?" Movement *always* gets our attention.

Obviously, still images don't "move" as video does. However, we can *imply* movement, even in a still image. This implied movement often makes for a more compelling image. (When we create video, movement becomes an effective tool for attracting and guiding the eye of the viewer. I'll cover this in Section 3, "Persuasive Moving Images.")

Movement is hard to show in a book. For example, if **FIGURE 2.3** were a film, your eye would go first to the lead horse because it is moving and bigger. Here in the book, the image *implies* movement. A great deal of commercial photography—especially fashion—uses implied movement to attract the eye.

Next, our eye goes to that which is in focus. If everything in the image is in focus, which happens with most smartphone shots, the eye skips to the next lower priority. But, if the focus varies within the image, our eye sees the object that is in focus before anything else, other than movement. This is why so many ads and movies present images where only a small portion of the frame is in focus. It tells the eye where it should look first.

FIGURE 2.4 illustrates this. The frame is filled with people, but we see the laughing woman first because she is the only one in focus.

FIGURE 2.3 Movement. The eye is drawn to movement or, in the case of a still image, the *illusion* of movement. (Image Credit: pexels.com)

FIGURE 2.4 Focus. Why does your eye see the smiling woman first? Because focus is more important than brightness or position. (Image Credit: Rene Asmussen / pexels.com)

Next, the eye goes to that which is different. This is why the black puzzle piece in Figure 2.2 attracted your attention. The entire image was white *except* for that one piece. Different could be a different gender or color or shape or…well, just about anything. Our eye instantly spots something that is out of place, or different, from the surroundings.

In **FIGURE 2.5**, you saw the rectangular color palette first because every other shape in the image was either curved or lighter in color. In other words, the simple *difference* of the color palette was enough to attract our attention first.

Continuing down our list of priorities, if there's no movement, everything is in focus, and all the elements are similar, our eye goes to that which is brighter. This explains why so many headlines in digital images are white. White attracts the eye, as you can see in **FIGURE 2.6**. (So does black against a white background, as in Figure 2.2, but that's because black is different compared to the white background, and difference ranks higher.)

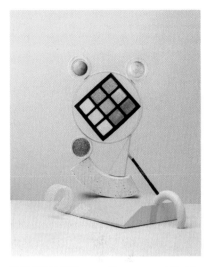

FIGURE 2.5 **Difference.** Your eye went first to the square of colors because it is different, in shape, color, and size, than anything else in the frame. (Image Credit: pexels.com)

In fifth position is size; the eye tends to look at the biggest element in the frame first. In **FIGURE 2.7**, as the woman looks back at the camera, she is the largest element in the frame. (She is also in focus, which also helps guide the eye.) We see her first and then look over her shoulder into the room to see what the meeting is about.

FIGURE 2.6 **Brighter.** Your eye went first to the woman on the left, because she is brighter than anything else in the frame.

FIGURE 2.7 **Bigger.** Your eye went first to the woman on the left, because she is bigger than anything else in the frame. (Image Credit: Elle Hughes / pexels.com)

FIGURE 2.8 **In front.** Your eye went to the man in front. When other criteria are missing, equal or not relevant, your eye goes to the subject in front. (Image Credit: Clarita Alave / pexels.com)

It is common to combine multiple priorities in a single image to make sure that the eye goes where you want. It is the rare image where only one priority is used.

In the end, if all else is equal, our eye goes to the object in front. In **FIGURE 2.8**, all the men are roughly the same height and size. They are all wearing similar clothes. So, where does our eye go first? To the person in the front.

The Six Priorities are really helpful in understanding how to catch and control the eye of the viewer. The eye doesn't stop exploring an image after its first look; rather, it explores the image based on the Six Priorities. By designing your image and text with these in mind, you can guide the viewer to see what you want them to see in the order you want them to see it.

Elements of the Six Priorities are rarely used by themselves; combining multiple priorities in the same image drives home where the eye needs to look first. We frequently use a "V" shape when positioning (called "blocking") actors in dramatic or dance scenes, as you can see in **FIGURE 2.9**. Positioning the lead actor or singer at the front of the V, combined with a slightly different—and, often, brighter—costume and increased lighting guarantees that the audience will focus on the person the director wants you to watch, in this case, the lead singer.

FIGURE 2.9 These concepts can be combined. Your eye *must* go first to the woman in front, because she is also brighter and bigger than any other person in the shot. (Image Credit: Cottonbro / pexels.com)

Our number one goal is to capture and retain the eye of the viewer so that we can deliver our message. Applying the Six Priorities can help you better control where the viewer's eye will look first, then second, and then third as it checks down elements on the list.

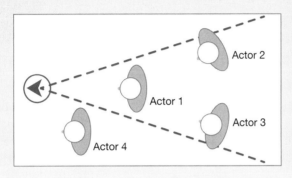
What Makes an Image Compelling

In 1991, Molly Bang, an award-winning children's book illustrator, published a trail-blazing book, called *How Pictures Work*, analyzing how the structure of an image works to engage the emotions of a viewer and how the elements of an artwork can give it the power to tell a story. If you can find a copy in your local library, it is worth spending a couple of hours reading it, because you'll spend the next year reflecting on it.

In her book, Molly investigated how changing shapes, positions, and colors would affect the visual telling of *Little Red Riding Hood*. In **FIGURE 2.10**, we see a visual representation of Little Red Riding Hood in the middle of the forest, about to meet the wolf. However, while the image evokes the story of Red alone in the forest, it isn't particularly scary. How could she change the composition of the image to make Red meeting the Wolf scarier?

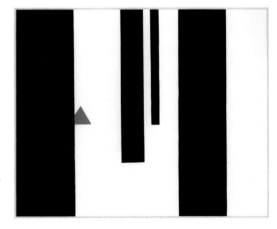

FIGURE 2.10 Using simple shapes cut from construction paper, Molly Bang experimented with the best way to tell the story of *Little Red Riding Hood* visually. (From *Picture This*, ©2000 Molly Bang. Used with permission from Chronicle Books, LLC)

Do not worry about whether the picture is pretty. Worry about whether it is effective.

—MOLLY BANG

As a result of her work, Molly discovered 12 principles that detail how the structure of an image affects its effectiveness. The first four describe the effects of gravity on how we perceive an image; the last eight reflect the world of the image itself.

1. Smooth, flat, horizontal shapes give us a sense of stability and calm. In fact, smaller horizontal or horizontally oriented shapes within a picture can also be felt as islands of calm.

2. Vertical shapes are more exciting and more active. Vertical shapes rebel against the earth's gravity. They imply energy and a reaching toward the heights or the heavens. If a horizontal bar is placed across the top of a row of verticals, stability reigns again.

3. Diagonal shapes are dynamic because they imply motion or tension. We tend to read images diagonally from left to right, the same way Westerners read a page of text.

4. The upper half of a picture is a place of freedom, happiness, and power; objects placed in the upper half also often feel more "spiritual." The bottom half of a picture feels more threatened, heavier, sadder, or constrained; objects placed in the bottom half also feel more grounded.

5. The center of the page is the most effective "center of attention." It is the point of greatest attraction and stability. However, shifting an element off-center makes it more dynamic because the edges and corners of the image are the boundaries of the picture world. Everything stops at the edge.

6. In general, white or light backgrounds feel safer to us than dark backgrounds because we can see well during the day and only poorly at night.

7. We feel more scared looking at pointed shapes; we feel more secure or comforted looking at rounded shapes or curves.

8. The larger an object is in a picture, the stronger it feels.

9. We associate the same or similar colors much more strongly than we associate the same or similar shapes.

10. Regularity and irregularity—and their combinations—are powerful. Though perfect regularity and perfect chaos are both threatening.

11. We notice contrasts; put another way, contrast enables us to see.

12. The movement and import of the picture are determined as much by the space between the shapes as by the shapes themselves.

Molly concludes, "When we look at a picture, we know perfectly well that it's a picture and not the real thing, but we suspend disbelief. For the moment, the picture is 'real.' When we are just beginning [to design], we often constrain our pictures...we make all the elements fairly similar in size; we tend to use the center of the page and avoid the sides; we tend to divide the space

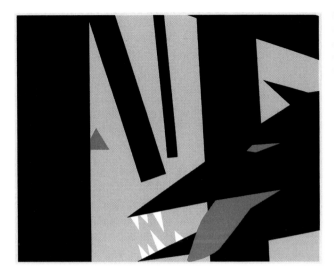

FIGURE 2.11 Again, simple shapes, but much more dynamic, exciting, and scary. Molly Bang's principles explain why. (From *Picture This*, ©2000 Molly Bang. Used with permission from Chronicle Books, LLC)

into regular sections; we tend to go for realism rather than essence....And very often we sacrifice emotional impact for 'prettiness.' Do not worry about whether the picture is pretty. Worry about whether it is effective."[2]

Molly Bang illustrates the power of these principles in **FIGURE 2.11**, which is strikingly more dramatic than Figure 2.10. As you'll learn, her principles extend our original Six Priorities. Let's explore these ideas further by looking at several specific visual composition techniques.

The camera represents the point of view of the audience.

Visual Composition Basics

An image contains more than content. Composition, blocking, and framing also play a role in how effectively an image communicates.

Changing Camera Angle: Wide vs. Close

Let's start with something simple: the angle of a shot. Wide shots show geography, where people and objects are positioned in the frame. We use these to set the scene. As the camera moves closer—and remember, as you move the camera, you are moving the audience—we see more detail. However, and this is important, close-ups also magnify emotion. The closer you get, the more emotion the shot contains.

Clearly, the most emotion is concentrated in the face, but because close-ups magnify even the smallest details—a twitch of a half-covered finger, a shoe moving stealthily across a carpet, or a drop of water hitting a surface—all contain significant emotional impact, depending upon the story.

[2] (Bang, 2016, pp. 52–92)

FIGURE 2.12 Compare the emotional difference between this wide shot... (Image Credit: Robert Stokoe / pexels.com)

FIGURE 2.13 ...and this close-up. The wide shot shows the sweep of her movements, while the close-up shows detail and emotion. (Image Credit: Robert Stokoe / pexels.com)

FIGURE 2.14 The wide shot (left) emphasizes the isolation she feels, while the close-up (right) explains why. (Image Credit: pexels.com (left), Engin Akyurt / pexels.com (right))

FIGURE 2.12 presents the environment where she is dancing, the sweep of her movement, and the reactions of the dancers behind her. The close-up in FIGURE 2.13, though, showcases the details in her costume and allows us to see the concentration and grace in her face.

In addition, the camera angle itself conveys emotion. Look at the combination of wide and tight shots in FIGURE 2.14. The wide shot sets the scene of her isolation, while the close-up explains why. Neither of these shots is "better." Rather, we need to understand their differences and decide which angle works best for our story.

Just as a story needs variations in pacing, so, too, do images. Generally, we tell a visual story by starting wide to show the audience where we are and then gradually move closer to amplify the detail and emotions in our story. Think of getting closer as an "emotional magnifier." You don't want to stay wide or tight all the time. The key is to shift position as your story evolves.

I also want to point out that getting extremely close to someone's face emphasizes the emotions in the image to such an extent that the person appears unattractive. You'll see extreme close-ups used a lot when we want to make someone seem out of control, out of touch, or generally unpleasant. Extreme close-ups are also often displayed in black-and-white to create a starker look. We'll discuss color in Chapter 5, "Persuasive Colors."

Changing Camera Position: High, Low, and Eye Level

Let's continue moving the camera, but rather than moving it closer or farther from our subject, let's move it vertically. Changing the height of the camera changes both where the eye goes first in the image and the emotional feeling of a shot.

The height of the camera can help control where the viewer's eye goes first. For example, in **FIGURE 2.15**, the height of the camera shifts the viewer's eye first to the background. Lowering the camera's height, shown in **FIGURE 2.16**, shifts the viewer's eye to the foreground first. Filmmakers use this trick often, starting with a low angle to identify the subject and then raising the camera to show us the environment they are moving through.

There's another emotional component to the height of the camera. If you think about it, when we were small children, everything around us seemed very big. We were essentially powerless, overwhelmed by our surroundings. As we grew and gained strength, many things that used to tower over us became smaller as we became stronger and more able to take care of ourselves. Now, when we talk with a friend, we try to meet them eye to eye, on the same level as they are.

While there is a stylistic difference between physically moving the camera closer to a subject versus zooming in, called "depth of field," the emotional context remains the same. We'll discuss depth of field later in this chapter.

HIGH ANGLES The camera is raised to emphasize the background. Cranes and tall buildings are often used to get the camera up high enough.

LOW ANGLES The camera is lowered to emphasize the foreground. Placing the camera a few inches off the ground puts extreme emphasis on the foreground.

FIGURE 2.15 High angles emphasize elements in the background. Your eye goes to the background first. The brightness of the hidden sun also helps attract the eye to the background. (Image Credit: Trace Hudson / pexels.com)

FIGURE 2.16 While low angles emphasize elements in the foreground. Your eye goes to the foreground first. Here, the camera is about 2 feet off the ground. (Image Credit: pexels.com)

FIGURE 2.17 The height of the camera, looking down, diminishes the man. (Image Credit: Gerd Altmann / pexels.com)

FIGURE 2.18 The hero shot. The camera is lower than the eyes of the two subjects to make these two normal men seem larger than life. The audience is "looking up" to them. (Image Credit: Shannon Fagan / 123RF)

These emotional values that we experienced growing up translate into how we perceive images:

- When the camera is high, looking down, as in **FIGURE 2.17**, it diminishes the subject, making them appear small and weak.
- When the camera is low, looking up, as in **FIGURE 2.18**, it makes the subject look heroic, bigger than life.
- When the camera is at the same eye level, the subject seems like one of us, a peer (**FIGURE 2.19**).

You'll see these angles used constantly in commercials to guide our emotions because they are so ingrained in how we perceive the world.

A low angle looking up (Figure 2.18) is so popular, it even has its own name: the "hero shot." Virtually every commercial or publicity photo uses this angle because it evokes feelings in the audience that we are not worthy to be associated with something so superior. "Yet," the narrator intones, "for just a few dollars even you can associate with something this glorious."

OK, maybe that's a bit over the top. But this angle is powerful, and it works, reliably, every time. Watch for it—you'll see it everywhere.

The emotions attached to the angle from which we view something also factor into sculptures of famous figures, like Julius Caesar in **FIGURE 2.20**. They are put on pedestals to force us to "look up to them." (The fact that Julius Caesar is also carved larger than life doesn't hurt either; it makes him bigger.)

Where You Hold the Camera Affects the Audience

Many times, when we take a photograph, we hold the camera at a height that is comfortable for us. However, that also means we are unconsciously placing an emotional twist on all our images, especially if there's a big difference in height between the person holding the camera and the subject of the shot.

Remember, the camera is the eyes of the audience. When you take a picture, think about what you want the *audience* to see and feel. The more you think of the camera as "the eyes of the audience," the better your pictures will be.

FIGURE 2.19 Here the camera is at eye level with the subject. Notice how this makes both the subject and the audience seem to be peers, even though she is a CEO. (Image Credit: Rebrand Cities / pexels.com)

FIGURE 2.20 A statue of Julius Caesar. (Image Credit: Skitterphoto / pexels.com)

Placing the camera at the same height as the eyes of the subject (Figure 2.19) makes both audience and subject peers. Newscasters use this angle constantly because they want to appear to be "one of the folks." You'll also see politicians and business CEOs use this angle. It says, "While I may have a lot of power, really, I'm just the same as you."

Blocking: Placing People and Cameras

The technical term for positioning actors and cameras is "blocking." It's the choreography of placing elements within the frame to create an image. Just as we can adjust the position of an actor, we can adjust the position, movement, framing, and focus of a camera.

Three terms that we use constantly in blocking are "foreground," that which is closest to the camera; "background," that which is furthest from the camera; and "mid-ground," that which is between the foreground and background.

"Blocking" is a term that is used in theater, dance, video, and film. It describes the process of working out the specific positions of actors, dancers, crew, and cameras for a scene or shot.

FIGURE 2.21 In this image, the two women have equal "weight." Both are the same size and seen in half-profile because the camera is centered between them. (Image Credit: Fauxels / pexels.com)

FIGURE 2.22 Moving the camera to one side allows us to see one person more fully so we can focus on them. Also, our eyes follow their eye line to go from their faces to the phone. (Image Credit: Anastasiya Gepp / pexels.com)

Shooting from an angle is generally more interesting than shooting from the center. However, when you want all subjects to have the same "weight," shooting from the center is the best option.

EYE LINE The direction of where someone is looking in the frame.

There's a general rule in film that, for any given scene, the actor who gets the first close-up is the subject of that scene.

Look at the image in **FIGURE 2.21** of two women having a conversation. The camera is placed evenly between them—what I call a "center shot"—so that the audience can watch both of them. In fact, it's hard to tell, from this shot, who's the subject of this photo; both faces are equal size and somewhat in profile. The eye isn't sure who to look at first.

FIGURE 2.22 illustrates another concept: "eye line." The eyes of both women are focused on the phone. Our eyes follow where their eyes are looking—as though there is a "line" connecting their eyes to where they are looking—leading us to see something specific in the image.

Compare Figure 2.21 with Figure 2.22. Look at what happens as we move the camera to the side to look over the shoulder of one of the women. The couple is still talking, still engaged in a conversation, but now the audience is given a much clearer clue as to whom the scene is about. The audience concentrates on the face it can see. In Figure 2.21, we are shown where everyone is located, but it's emotionally neutral. The emotion gets a big boost by moving the camera to the side and moving in—called "going tighter"—for a closer, clearer shot.

The Fourth Wall

These two photos of the two women chatting are also a good illustration of the "fourth wall." This is the concept that the camera acts as a "wall" between what's in the shot and the audience looking into the shot. In other words, audience members are voyeurs, eavesdropping on scenes and conversations. We can break this wall, and I'll discuss how in the section on eye contact later in this chapter, but for now, it is important to note that the audience is *watching* this conversation, not *participating* in it.

The 180° Rule

Determining the position of the camera in relation to the talent—that is, the people or objects being photographed—is such an important part of photography that it has its own name: the "180° Rule."

The 180° Rule (**FIGURE 2.23**) illustrates where to put cameras to get the best wide shots and close-ups. For example, if you connect the noses of two people talking, the best close-ups come when the camera is placed close to the line, over the shoulder of the person opposite to speaker. When shooting video, the rule further states that, to avoid confusing the audience, all cameras must be on the same side of the line.

As we move the camera to the side, our ability to read faces quickly improves the image. The center of interest for that shot becomes much more obvious. There's a lesson here. Get out of the center and move to the edge. Your images will improve.

FIGURE 2.24 illustrates this. The camera on the left shoots the close-up of the woman on the right. The camera on the right shoots the close-up of the man on the left. The center camera provides the wide shot so we can keep track of everyone in the shot.

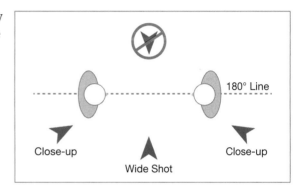

FIGURE 2.23 An illustration of the 180° Rule.

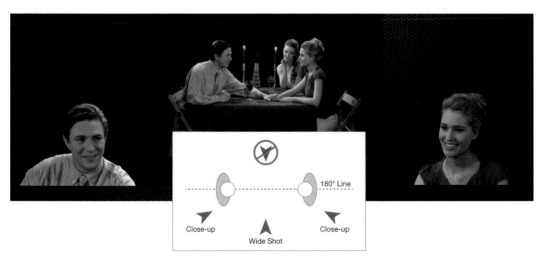

FIGURE 2.24 An example of the 180° Rule in action in video. (Images courtesy of 2ReelGuys.com)

FIGURE 2.25 Notice how boring this center wide shot is. (Image Credit: Fox / pexels.com)

When shooting stills, the key benefit of the 180° Rule is determining where to get good close-ups. When we get to video, however, this rule assumes a whole new level of importance, which we'll cover in later chapters.

In **FIGURE 2.25**, everyone is the same size. Each is equally remote from the camera. The eye doesn't know where to look first, except you probably looked at the computer in the foreground first, because it's the biggest subject in the frame. This is truly a boring shot.

Walking Toward the Camera

Another note about blocking. It is inherently more interesting to have a subject walk to or from the camera than to walk from one side to another, as you can see by comparing the two shots in **FIGURE 2.26.**

If a subject is walking to the camera, they are also walking into a close-up, which boosts the emotional content of the shot. If they are walking away, they are moving into a wide shot, which, though it minimizes the subject, maximizes our understanding of their relationship to the environment.

Also, in Figure 2.26, we see the woman in the black jacket first because she is different and big. She is also in the center of the frame. However, notice how quickly your eye gets bored and starts looking elsewhere in the frame for something else to look at. We'll talk about this more in a couple of pages when we discuss symmetry versus balance.

FIGURE 2.26 Walking to or from the camera (left) is much more visually interesting than walking from one side to another (right). (Image Credits: Just Name / pexels.com (left), Clem Onojeghuo / pexels.com (right))

Framing and the Rule of Thirds

Camera framing is how the frame is positioned around the elements in an image. We also use the term to describe how an element within the image is positioned relative to the frame. As we learned from Molly Bang, objects in the center of the frame have the most power, but objects at the edges are more interesting. What I've discovered is that our eyes get bored when a subject is placed dead-center. We quickly start exploring the rest of the frame to see what else there is to see. However, images placed toward the edge have a dynamic about them that keeps drawing the eye.

This observation isn't unique to me; the Dutch master painters in the 1600s discovered and popularized it through their paintings. Like the 180° Rule, this, too, has its own name: the "Rule of Thirds." This is another important concept that will recur throughout this book.

RULE OF THIRDS This divides the frame into thirds horizontally and vertically to describe a way of positioning important elements at intersections to achieve a dynamic visual balance.

The Rule of Thirds divides the screen into three equal rows and columns (**FIGURE 2.27**). Important elements of the image should be placed either at one of the four intersections or along one of the lines to create a pleasing, dynamic visual balance.

As you look at Figure 2.27, notice that the woman is not centered. In fact, she is looking "into" the frame, with her body positioned on the side of the frame opposite to where she is looking. This space between her eyes and the left or right edge of the frame is called "looking room."

LOOKING ROOM The distance between a subject's eyes and the edge of the frame. Reducing looking room increases a sense of limits or claustrophobia.

When I first presented the concept of the frame, I mentioned that it fools the brain into thinking that only that portion of the image that we see in the frame exists. Looking room reinforces this.

FIGURE 2.27 A shot framed according to the Rule of Thirds. Notice, also, that she is framed so she has lots of "looking room." (Image Credit: Mentatdgt / pexels.com)

Compare the emotional difference between Figure 2.27 and **FIGURE 2.28**. In real life, both are in a similar environment—looking out a train window. However, the image of the woman with lots of looking room feels "normal," while reducing the looking room makes the shot of the man feel like he is running out of options. The framing differences between these two shots vastly alters how we interpret the emotions in each shot.

HEAD ROOM The distance between the top of a subject and the top of the frame.

In addition to looking room, which adjusts the horizontal distance between the subject and the sides of the frame, there's also "head room," which is the distance between the top of a subject and the top of the frame. Here, too, the Rule of Thirds holds sway.

In these last two framing examples—Figures 2.29 and 2.30—did you see that both subjects were framed with looking room? The subject was moved slightly to the left side of the frame (called "panning") so they can "look into" the wider portion of the frame. Looking into the wider portion of the frame is considered normal framing.

In general, you want the central focus of the subject placed on the upper Rule of Thirds line. In the case of a close-up of a person's face, that means putting their eyes on the line.

FIGURE 2.29 illustrates a common problem with amateur photographs: too much, or too little, headroom. If the man is the subject of the photo, as opposed to the building behind him, put his eyes on the top line of the Rule of Thirds.

Another problem, shown in **FIGURE 2.30**, is putting the face of the subject in the center of the frame. In both shots, the man's head is the same size. Yet, look at how much more substantial and forceful the image on the right is. The only change was moving the eyes from the center of the frame to the top Rule of Thirds line.

FIGURE 2.29 Three examples of headroom: too much (left), too little (right), and just right (center). The white lines indicate the horizontal lines from the Rule of Thirds. (Image Credit: Cottonbro / pexels.com)

FIGURE 2.30 The size of his face is the same; all that's changed is moving the face from the center of the frame (left) to the top Rule of Thirds line (right). (Image Credit: Malcolm Garret / pexels.com)

Eye Contact Is Nontrivial

Earlier in this chapter, I discussed the concept of the "fourth wall." This is the assumption that the audience is invisible to the subjects in an image; they are "hiding behind a wall."

The images in FIGURE 2.31 make this concept vividly clear. In the left image, we are viewers. In the right image, we are participants. With the image on the left, the audience is invisible, watching a little boy play with a camera and a soccer ball. It's the type of scene to make a parent smile; because the kid is so caught up in what they are doing, they've tuned out everything around them.

However, the image on the right, where the boy's eyes connect directly with the camera, means the audience is no longer invisible; we are directly involved in what's happening in the scene.

FIGURE 2.31 Note the emotional difference between looking *into* the frame and being looked *at*. (Image Credit: Tuấn Kiệt Jr. / pexels.com)

FIGURE 2.32 Notice how much more pleasing the framing is if you don't cut someone at body joints. (Image Credit: Italo Melo / pexels.com)

You can't have talent looking at the camera in one shot and not looking at it in the next shot.

This is another important concept. You can't have talent looking at the camera in one shot and not looking at it in the next shot. It is disjointing to the audience. We don't know what our role is: Are we viewers or participants? Pick one option and stick with it. This popular, but unsettling, trend of bouncing between eye contact and no eye contact has diminished the impact of many interviews.

Cropping is the process of removing portions of the image, either to get rid of garbage in the shot or to alter the framing. When you crop an image (**FIGURE 2.32**), don't cut someone at body joints, meaning their neck, elbows, wrists, or waist. It is much more pleasing, as Figure 2.32 illustrates, to crop images elsewhere on the body.

Symmetry vs. Balance

There's a big difference between symmetry and balance. Symmetry is where both sides of an image are the same, as shown in **FIGURE 2.33**. Balance is where the "weight" of images on one side of the frame "balance" those on the other side. This picks up on ideas we first learned from Molly Bang, earlier in this chapter. Symmetry creates lovely images, but balance is more compelling.

In Figure 2.33, both sides of the image are symmetrical. It's lovely to look at, but our eyes quickly start looking for something else. There's too much emphasis on the center of the frame.

See how quickly you stopped looking at the architecture and shifted to the next image to check out the rocks?

We also saw this problem with symmetry in Figure 2.27, with the woman in the black jacket walking up from the beach. She was in the center of the frame, but our eye quickly got bored with the subject in the center.

FIGURE 2.34 These rocks, while off-center (though still framed according to the Rule of Thirds), make for a much more compelling image. (Image Credit: pexels.com)

FIGURE 2.33 The symmetry in this photo creates a stunning visual, but, the eye quickly gets bored and looks elsewhere. (Image Credit: pexels.com)

In **FIGURE 2.34**, balance combines with the Rule of Thirds to create a much more compelling image. We see the rocks first, because they are bigger. Then, we start exploring elsewhere in the frame, but we keep coming back to the stack of rocks. That's the power of the balance—creating dynamic images that capture the eye while allowing it to explore the frame in more detail.

As another example of balance, the tree image in **FIGURE 2.35** has a number of elements going for it: The tree is brighter than the background, is bigger than other trees near it, is framed according to the Rule of Thirds, and contains different colors when compared to the background. The eye is automatically attracted to it! (The streaks of sunlight serving to backlight it help, too.)

FIGURE 2.35 Another example of balance, with the brightness of the tree balanced by the heavier, darker woods on the left. (Image Credit: Ray Bilcliff / pexels.com)

FIGURE 2.36 Here's the same photo. Flipping the image changed the meaning. (Image Credit: Marius Venter /pexels.com)

The Weirdness of Right-Handedness

There's a fascinating psychological quirk: right-handedness. Readers of Western languages also tend to "read" images from left to right—the same as a book. This means that our eyes give more weight to where they finish looking (the right side) than where they start (on the left). This often affects how we interpret images.

In **FIGURE 2.36** the photo was flipped horizontally. When it did, the meaning changed. In the left image, we see the opening first; in the right image, we see the man first. Why? Because in both cases, our eye gave more weight to the right side of the image, then we used the right side of the image to interpret the meaning of the complete image.

Depth of Field

Depth of field is principally determined by the lens angle combined with its aperture. Wide shots tend to have a deep depth of field, while close-ups, especially when using a zoom lens, have a shallow depth of field.

The last concept I want to illustrate in this chapter is one of the most important: depth of field. This determines which parts of the image are in focus. Focus, more than any other element aside from movement, determines where the eye will look first.

At the beginning of this chapter, when I listed the Six Priorities, focus was second. Our eye goes to that which is in focus, regardless of where it is located in the frame. Focus, when you can control it, is a great way to make sure the eye sees the subject you want it to see.

There are two types of depth of field: deep and shallow. Deep depth of field allows almost all of the image to be in focus, while shallow means that only a limited portion of the image is in focus.

FIGURE 2.37 This illustrates deep depth of field. Almost everything in this image is in focus. Most cell phones take this type of shot. (Image Credit: pexels.com)

FIGURE 2.38 In this shot, the depth of field is so shallow that only her face is in focus. Both the sparklers and the background are soft. (Image Credit: Renato Abati / pexels.com)

FIGURES 2.37 and **2.38** illustrate the differences between deep and shallow depth of field. We use depth of field to draw the eye to the important elements in an image. In Figure 2.38, the photographer focused on her face. She could just as easily have put the sparklers in focus, with everything else soft. But, she chose to focus on her smile, rather than what was making her smile.

Does Sex Sell?

Before we close this chapter, we need to talk about the 800-pound gorilla of marketing: sex. *(Cue dramatic music, with red arrows pointing to* **FIGURE 2.39.***)*

Hooking sex to products has a long history. Starting with Pearl Tobacco (1871) and continuing with Woodbury's Facial Soap, Jovan Musk Oil, Benetton, Carl's Jr., Victoria's Secret, and the poster child for sex in advertising, Calvin Klein jeans, each of these company's fortunes turned from poor to profitable by painting a direct line from their products to sex (see sidebar, page 49).

Still, does sex *sell*? In a word: maybe. Even today, with sex in advertising seemingly everywhere, there's no guarantee that it works.

FIGURE 2.39 From romance novels to laundry detergent, sex is everywhere—but does it sell? (Image Credit: Majdanski/Shutterstock)

Matt Flowers, CEO of Ethos Copyrighting, wrote in his blog that if your audience is interested in sex, "Remember that they have access to that content with just the click of a button. So you have to ask yourself, 'why try to compete with the plethora of online sites that make sex their sole business?'"[3]

According to Magda Kay, writing in *Psychology Today*, "If you ever wondered whether using sex in advertising helps to sell, here is the answer: it does. Actually, [sex] is one of the strongest and most effective selling tools. The relationship between sex and marketing is a winning combination for almost any business. However, if you don't know how to use it, you're risking putting off your potential customers."[4]

According to *Psychology Today*, a literature survey in 2017 by Wirtz, Sparks, and Zimbres sought to determine the effects of sexually provocative ads. They assessed effectiveness in terms of:

- How well the ads were recalled
- Whether they generated a positive or negative reaction
- How the sexually oriented ad compared to a similar ad without the innuendo
- What people thought about the brand
- Whether the ad actually generated sales

The results were interesting. In summary:

- In terms of capturing attention, sex works. However, people remembered the *ad* more than the *brand*. "While sex might be attention-grabbing, it didn't seem especially good at getting people to remember the objects being sold."
- Regarding people's attitude toward the ads, they were modestly positive for men and modestly negative for women. In other words, it was essentially a wash.
- Curiously, both men and women expressed negative feelings, on the whole, for brands that used sex to sell things.
- When it came to purchase intentions, it seemed that sex didn't really sell, but it didn't really seem to hurt, either.

"While it might be useful for getting eyes on your advertisement, sex is by no means guaranteed to ensure that people like what they see once you have their attention. In that regard, sex—like any other advertising tool—needs to be used selectively, targeting the correct audience in the correct context if it's going to succeed at increasing people's interest in buying. Sex in general doesn't sell."[5]

[3] (Flowers, 2018)
[4] http://psychology
formarketers.com/
sex-and-marketing/
[5] https://www.
psychologytoday.com/us/
blog/pop-psych/201706/
understanding-sex-in-
advertising

This ad, from Woodbury's Facial Soap in 1916, is considered the first mainstream use of sex in advertising. Not surprisingly, the first brands to use sex in their ads were saloons, fashion, tonics, and tobacco. The earliest known use of sex in advertising was in 1871 when Pearl Tobacco placed a lightly draped female figure rising from the waves on its label.[6]

In a 2015 study, authors Robert B. Lull and Brad J. Bushman of Ohio State University wrote, "Brands advertised using sexual ads were evaluated less favorably than brands advertised using nonviolent, nonsexual ads. There were no significant effects of sexual ads on memory or buying intentions. ...As intensity of sexual ad content increased, memory, attitudes, and buying intentions decreased."[7]

Sex gets attention but doesn't always drive behavior.

I think a good way to summarize these findings is that sex gets attention but doesn't always drive behavior. Also, keep in mind that sexy ads tend to go hand in hand with objectification. Even today, with raised awareness caused by the #MeToo movement, women are still five times more likely to be shown in revealing clothing than men. As well, sexy ads tend to reinforce stereotypes. "Recent 2017 research from the Geena Davis Institute on Gender in Media," Matt Flowers wrote, "found that women were 48% more likely to be shown in the kitchen than their male counterparts. Moreover, men are 89% more likely to be depicted as smart in comparison to women."

Matt Flowers continued, "Sexual content in the age of digital marketing is alive and well as far as organic content is concerned, but you're not going to run a pay-per-click campaign with that strategy. While innuendos can be funny, suggestive themes can be stylish, and sex can sell, it's important that marketers consider their audience carefully so as not to offend potential customers nor be provocative for the sake of publicity.

[6] (1916 Ladies' Home Journal version of the famous seduction-based ad by Helen Lansdowne Resor at J. Walter Thompson Agency. Public Domain.)
[7] (Lull, Bushman, Brad J., 2015)

"While using sex in advertising can be a successful tool for drumming up sales, it can also have negative consequences for both your reputation and our culture. For that reason, get to know your brand, what it stands for, and who your audience is."[8]

Sex is like any other image. Used with purpose and with an understanding of your message and audience, it can work. Used simply to attract the eye without connecting it to your product or message means your visuals join the thousands of others that our eyes glaze over each day.

KEY POINTS

We covered a lot of ground in this chapter, and we'll return to these concepts many times throughout this book. Here is what I want you to remember:

- The camera represents the point of view of the audience.
- Images, even still images, tell stories and evoke emotional responses in the viewer.
- The eye looks at, and explores, an image based on the Six Priorities of where the eye looks first.
- The mind assumes, when it sees a frame, that there is nothing outside the frame.
- Blocking is the process of positioning elements and the camera to create the best image for the story you want to tell.
- Framing is the process of determining where the frame is placed around the image.
- There are many techniques we can use when we capture or create an image that will enhance the story we are telling and generate an emotional response in the audience.

PRACTICE PERSUASION

Using your cell phone or camera, take some photos of a family member or friend. Alter the position of the camera and the framing based on the ideas in the chapter.

Look at the images and see whether they convey different emotions or perspectives to you. Then, show them to someone else and see whether they get the same or different reactions.

[8] (Flowers, 2018)

Just as a note, don't ask the person you photographed to critique your photos. They will be too busy worrying about how they look to pay any attention to your photography. For this reason, most talent are not good judges of the programs they are in.

PERSUASION P-O-V

IDEAGRAMS

Dennis Starkovich
Ideagram.co
Website: www.ideagram.co

In 1979, when I was 30 years old, I started creating diagrams of an analytic and personal nature that were an outgrowth of my professional role as a management consultant. I specialized in providing PERT/CPM network analysis for planning, developing, and implementing projects of all types, sizes, and complexity. I realized that this technique could also apply in our personal lives.

This was my initial insight into creating "Idea-Grams." These took the PERT/CPM methodology of process and project planning into a tool for examining written material and extracting the core elements or ideas into a visual diagram.

I consider an IdeaGram to be a mix of alchemy and artistry for ideas. An IdeaGram (which is my term) is a unique, evolutionary approach to visual diagramming to analyze and convey the intrinsic ideas, content, and wisdom contained in written or narrative material. Mind maps, which you may have heard about, and IdeaGrams are complementary in that the mind map is first and foremost a brainstorming tool while the IdeaGram chart is best suited for analyzing and presenting ideas in a methodical, sequential, and inherently legible manner.

The unique aspect of the IdeaGram charts is that they succinctly and graphically extract the core ideas from a narrative, in a visually sequential and legible manner.

The reference to alchemy is an acknowledgment that concentrating on ideas in this way helps initiate a change in one's understanding, perspective, and perception. The concentration required to study and extract the information from a written narrative is unique, somewhat difficult to perform, but exceedingly rewarding to undertake and accomplish.

The reference to artistry is that the presentation to an audience of an Idea-Gram prepared by another person provides both a view into the mental framework and understanding of the person who created the IdeaGram and an illustration of the thought that had been essential to development of the initial written narrative.

Who could use IdeaGrams? My answer is, everybody! IdeaGrams have unlimited creative as well as professional application. They are useful in areas of new learning, self-improvement, planning, time management, and education. They provide us with a new model for creatively mapping any project, goal, or understanding.

I'm sincerely interested in sharing the clarity of thinking that develops as a result of applying the techniques of critical analysis and concentration that are required to analyze a subject or narrative and create a chart that reflects the content of the process or written material.

The illustration is a sample IdeaGram, in this case to explain the process itself.

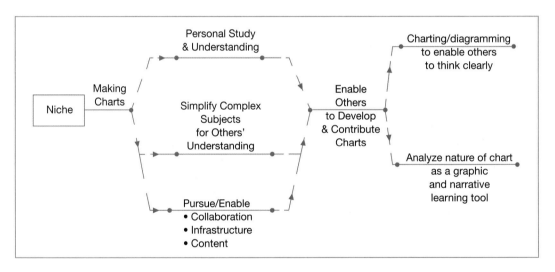

Here's how I define these different steps:

- **Personal Study and Understanding.** The development of these charts and diagrams over the past 40 years stemmed from a personal vision that sometimes resulted from the necessity of simplifying a complex series of ideas into a simple visual illustration. At other times, the impetus was simply to capture what might have been a complete book into a simple chart.

- **Simplify Complex Subjects for Others' Understanding.** This began as I was raising a child and observing and participating in the development of his knowledge growth from kindergarten to adulthood. Having encountered meaningful concepts that I feel others might appreciate, I began a process of categorizing and exhibiting them as IdeaGrams.

- **Collaboration.** Develop a platform for expanding the "library" of Idea-Grams and enable others to share their creations and subject matter. Similar sites exist for "mind maps," but mind maps, in my opinion, are not easily read (i.e., left to right, top to bottom, visually organized in logic and categorization).

- **Infrastructure.** Develop a tool that digitally enables the creation of IdeaGrams in the diagramming format shown.

- **Content.** Implement the creation of a significant volume of educational material to help others of all ages to understand, experience, and create similar documents and expand their understanding of various subjects.

- **Enable Others to Think Clearly.** I believe that the clarity of thinking that is required to create an IdeaGram can be learned. This clarity develops as a direct result of the critical thinking and concentration that are required to analyze a subject and create a chart that reflects the content of a process or a piece of written material.

- **Charts Become a Graphic Learning Tool.** Even as a picture is worth a thousand words, there is something special about an IdeaGram that conveys special insight into a process or narrative. I seek a wider discussion of this concept. I know what the development and use of this process have contributed to my awareness, understanding, and patience regarding my observations. I would like to better understand what effect this tool would have with others as they learn, explore, and collaborate with the use of this tool.

As I write this description for *Techniques of Visual Persuasion*, I am more and more interested in sharing these techniques with others. I have never stopped reading, learning, studying, diagramming, and applying the process to my personal knowledge, understanding, and development. I invite you to join me on this journey!

If St. Paul had been talking to a classroom of prospective writers, he might have changed a consonant. Faith and hope must abide, for the writer must have faith in his own ideas, and he must hope that his words will be read, but after faith and hope comes clarity. Without clarity we are not even sounding brass or tinkling cymbals. Be clear, be clear, be clear! Your idea or image may be murky, but do not write murkily about it. Be murky clearly!

JAMES J. KILPATRICK[1]
JOURNALIST, AUTHOR, AND GRAMMARIAN

CHAPTER 3

PERSUASIVE WRITING

The goals of this chapter are to:

- Explain the importance of finding and crafting the right message
- Explain the role of storytelling
- Define a creative workflow
- Describe how a workflow improves efficiency
- Explain how to plan a story using an "elevator pitch"
- Provide tips for writing smarter and more powerfully

[1] (In 1 Corinthians 13:13, St. Paul talks about "faith, hope, and charity." This quote paraphrases that Bible passage. Kilpatrick, 1984, p. 32)

AS I WAS RESEARCHING this chapter, I came across *Write Better, Speak Better*, a Reader's Digest book written in 1972, before social media, before email, even before personal computers. Yet, its thoughts are still relevant.

"Let's look carefully at this matter of being *genuinely persuasive*.... Inducing a person to change his mind is a very delicate operation indeed.... Don't try to bludgeon your listeners into submission with facts, figures and what debaters like to refer to as 'conclusive proofs.' Remember, your listener will accept your idea only if he first *wants* to accept it."[2]

Persuasion is something we do to create a choice for the viewer by developing a focused message, wrapped in a story and delivered with emotion to the right audience.

The Right Message Is the First Step

In the cartoon for **FIGURE 3.1** the two women had the best of intentions. They had a compelling topic and forceful talking points. They were targeting the right audience. But they weren't connecting. Why? Because they assumed their audience was driven by the same concerns they were. The women were focused on what *they* felt was important, and, in so doing, they lost sight of what their *audience* felt was important. Once they took a step back and looked at the problem through the eyes of their audience, they instantly and successfully connected.

Persuasion is something we do to create a choice for the viewer. That was true 50 years ago and remains true today. So, how do we set up this choice? By developing a focused message, wrapped in a story and delivered with emotion to the right audience.

It may seem strange, in a book on visual techniques, to devote a chapter to storytelling. But, stories are deeply woven into the human psyche. Stories came first. Sitting around a fire telling stories is as old as humanity itself. Stories provide a direct line to the brain and the heart.

FIGURE 3.1 Right audience, wrong initial message. Focus on the needs of your *audience.* (Image Credit: Pearls before Swine by Stephan Pastis)

PEARLS BEFORE SWINE By Stephan Pastis

[2] (Reader's Digest, 1977, pp. 528–536)

Even in today's image-oriented world, words matter. Need proof? Look no further than today's memes.

Memes like **FIGURE 3.2** combine the power of words with the visceral hook of an image. The combination of words and image is far more powerful than just words or image alone.

FIGURE 3.2 Memes combine the power of words with the eye-grabbing hook of an image. (Image Credit: Ali Pazani / pexels.com)

The basic definition of a story is: "A hero struggles to overcome obstacles to reach an important goal." Or, as I tell my students: A story takes an audience on a journey, encompassing challenges and change, structured with a beginning, middle, and end.

Stories are not just a report. "The purpose of the report is to *point* you there," writes Roy Peter Clark, in his book *How to Write Short,* "but the story *puts* you there."[3] I really like his description of *reading* a report but *living* a story. Stories are involving and compelling, capturing the imagination. Good stories touch your emotions and make you feel. That, not surprisingly, is also true for images.

As famous film director Alfred Hitchcock once said, "There are three elements that make a film great: the script, the script, and the script."

This raises an important question: What makes a story good? This is like asking, "What's the best vehicle?" A school bus, pickup truck, or sports car are all "best," depending upon your task. They can also be the worst, if the task changes. (The image of a sports car carrying plywood came instantly to mind.)

The only answer to "What's a good story?" is that it depends upon what you need it to do.

We read reports, but live stories. Stories touch our emotions and make us feel.

Images Evoke Stories

Even if no story is provided, our brains automatically invent one for an image. We don't simply look at a photo and say, "Oh! That's a photo." Instead, we look at it and wonder, "What's going on? Who is that? Why is this image important?" Without much effort, a story appears in our imagination.

Since this story creation process occurs in our viewers whether we lead them there or not, it is much better for us to suggest the story than have them invent one on their own.

[3] (Clark, 2013, p. 67)

FIGURE 3.3 Our brain invents stories when it looks at an image. (Image Credit: pexels.com)

If we don't provide a story to go with our images, the viewer will create a story of their own. We need to control this conversation, not the viewer.

For example, our eye goes first to the firefighter in **FIGURE 3.3** because, according to the Six Priorities, he's bigger, different, and the most in focus element in this shot. (He's also framed according to the Rule of Thirds.) As we look at this dramatic action photo, we want to know more: What fire, why is he shouting, and who is he shouting at? Already, our minds are inventing a story to go with the image.

Each image begs us to create a story to explain it. We want to know what happens next. I'm not saying that posed pictures should be avoided. Rather, when we are telling stories, the eye and brain are more engaged by something dynamic, something moving, or something changing than by a lovely static image.

The key point is: If we don't provide a story to go with our images, the viewer will create a story on their own. It is far better for *us* to control this conversation than the viewer.

Stories involve change; it's essential to any story. I illustrate this in my class by inviting one of my students to walk to the front of the room and sit on the edge of a table. All the students watch as she walks up and sits. After a few seconds of the student just sitting there, the class turns their eyes back to me, wondering what will happen next. I say nothing, and after a few more seconds, everyone starts to fidget.

At this point, I ask the student sitting at the front to stand, walk to the wall, turn, walk back, and sit on the table again. Everyone watches as the student does this. At that point, I say, "This illustrates the power of change in your stories. If nothing is happening, you lose the audience. As soon as things start changing, for good or bad, your audience is right back with you."

In any story, change is essential.

Since we are trying to persuade, that is, to create change through our visuals, these two elements—change and stories—go hand in hand with what we are trying to accomplish.

Get Organized with a Workflow

As we get ourselves organized for storytelling, let me first define some key terms that apply to persuasive storytelling:

- **Theme.** Also called the "elevator pitch." This is a short description, only two or three sentences, of what you want to convey and what you want the audience to do with that information.

- **Goals.** Measurable objectives that will be used to determine whether your messages have any impact. For example, decrease the amount of street trash by 10 percent.

- **Audience.** The group of people who are the target of your message. A story can exist only in terms of its audience. For example, elementary students, high-school students, and local businesses would each have different stories.

- **Message.** The *content* of what you want to tell the audience. For example, "Pick up your trash."

- **Story.** The *form* of how the message is told for each audience. A single message could, and generally does, have multiple stories associated with it. For example, a story for grade-school children would show them where to throw out trash. A story for high-school kids would ask them to volunteer to pick up litter. A story for businesspeople would ask them to provide more recycling opportunities.

- **Call to Action (CTA).** An explicit action you want the audience to take after seeing your story.

I discovered that I often use "story" and "message" interchangeably, but they really aren't. The message is the *content* you want to convey. The story is *how* that content is conveyed. For example, let's say the message is: "Don't pollute." Stories could be images of recycling or picking up litter or a polluted landscape. In other words, there are multiple stories that can all revolve around the same message.

WORKFLOW An explicit, usually written, series of steps for completing a task efficiently.

Efficiency is a theme that recurs throughout this book. One way to boost efficiency is to create a *workflow*. This is an explicit, usually written, series of steps (called a "process") for completing a task efficiently. Creating stories for visual persuasion is similar to building widgets. Both are done often, and both consist of repetitive tasks.

In today's world there just isn't enough time for us to get our work done to our satisfaction. As I tell my students, "Do the best you can in the time you've got." Jan Fedorenko, senior media manager at WLS-TV, shared her mantra: "It's better than good; it's done!" We need to be efficient—*really* efficient—with the time we have.

I've learned that the best way to accomplish any repetitive task is to create a workflow for it. By following a workflow, we stay focused on the job at hand and don't waste time reinventing the wheel.

Creative projects that aren't repetitive, such as the first time you do something, will benefit from a workflow. However, they will benefit even more from time spent planning. We'll cover more about planning later in this book.

In a saying first attributed to the poet Paul Valéry, any creative work is never finished—it's abandoned when time runs out. While, creatively, we can always make our work better, at some point there just isn't any more time. That's why efficiency is so important.

For example, here's a workflow you can use to create a persuasive story:

1   Create a theme describing what you want to accomplish.

2   Set measurable goals for that theme.

3   Define your audience.

4   Develop a message targeted at that audience.

5   Create a compelling story from that message.

6   Create an image that illustrates the story.

7   Create a Call to Action that pays off the story.

Don't waste time doing something now that you plan to change later.

There's no sense working on an image if you haven't figured out the audience yet. Don't waste time doing something now that you know you will change later in the process. For example, with this workflow, you can concentrate on each storytelling task, complete it, then tackle the next task, without retracing your steps. A workflow provides a map outlining the steps necessary to accomplish a task so you can concentrate on the creative process.

As we saw in Chapter 2, "Persuasive Visuals," there are many different ways of telling the same story. For example, in the story of *Little Red Riding Hood*, Molly Bang used construction paper cutouts. Michael Bay would have used a movie camera, while Picasso would have used a paintbrush.

The Essence of Persuasion

According to Aristotle, who created the entire field of persuasion, called "rhetoric," there are three elements of persuasion:

- The character of the speaker (*ethos*)
- The emotional state of the hearer (*pathos*)
- The persuasive argument itself (*logos*)[4]

In other words, is the speaker credible, is the audience in an emotional state that allows them to receive the message, and is the message itself compelling?

The four pillars of trust are reliability, capability, honesty, and empathy.

As we first learned in Chapter 1, "The Power of Persuasion," what makes persuasion successful is not cleverness, but trust. The viewer has to trust you. "...trust is not only the irreplaceable basis of all communication but is also one of the more profound channels of connection between human beings."[5] The problem is that in today's world, trust is in very short supply. Your audience automatically assumes you are not telling the truth.

Phillip Khan-Panni describes the "four pillars of trust" as reliability, capability, honesty, and empathy.[6] Be careful taking liberties with the truth—even in jest. In today's connected society, it is far too easy to be found out, and the results won't be pretty. During the course of this book, you'll learn how to create, modify, and distribute images. However, to quote the iconic line from Uncle Ben in Spider-Man, "With great power comes great responsibility." Once you lose the trust of your audience, you will have a hard time getting it back.

Examples of stretching the truth abound. I began writing this book in the heat of the 2020 American political season. I completed it in lockdown in California during the COVID-19 pandemic. The lack of trust and misinformation surrounding both were pervasive and debilitating. Political ads have refined image manipulation into an art form. Politicians learned long ago that images are far more powerful than policy statements. The problem is that excessive manipulation and exaggeration have also destroyed the trust we need to have in our political institutions. Gaining it back will take a long, long time. There's a lesson for us and our message amid all this excess as well.

[4] https://plato.stanford.edu/entries/aristotle-rhetoric/#4.1
[5] (Bridges & Rickenbacker, 1992, p. 8)
[6] (Khan-Panni, 2012, pp. 80–83)

"Persuasion must be sincere."[7] Be careful of using fancy words when a simpler one would work just as well. George Orwell, in his book *Politics and the English Language*, wrote, "The great enemy of clear language is insincerity. When there is a gap between one's real and one's declared aims, one turns as it were instinctively to long words and exhausted idioms, like a cuttlefish spurting out ink.... Never use a long word when a short word will do."[8]

Interesting stories have focus, wit, and polish. Like good poetry, no words are wasted; each word needs to carry its own weight. Use these three questions to focus your story:

- Why does this story matter to my audience?
- Why does this story need to be told?
- What's the point of this story?

Start with the Basics: An Elevator Pitch

The easiest way to plan is to start small. I suggest creating an "elevator pitch," which is also called an "elevator speech." Terri Sjodin is the principal and founder of Sjodin Communications, a public speaking, sales training, and consulting firm. She devoted an entire book, *Small Message, Big Impact*, to creating and presenting an elevator speech.

Chapter 1 stressed that we need to create a clear message. Clarity is at the heart of any elevator speech. As Terry writes, "An elevator speech is a brief presentation that introduces a product, service, philosophy, or idea. The name suggests the notion that the message should be delivered in the time span of an elevator ride, up to about three minutes. Its general purpose is to intrigue and inspire a listener to want to hear more of the presenter's complete proposition in the near future."[9]

Who Coined the Term "Elevator Pitch"?

According to Wikipedia, there are at least three possible origin stories, but the one I like is that in the late 1960s, Philip Crosby, a quality test technician at ITT, wanted to propose a change at the company. He waited at the ITT headquarters elevator until the CEO arrived. Crosby stepped onto an elevator with the CEO to deliver his pitch. Once they reached the floor in which the CEO was getting off, Crosby was asked to deliver a full presentation on the topic at a meeting for all of the general managers. Out of that presentation came a promotion and his first book, *The Art of Getting Your Own Sweet Way*, and, ultimately, his own consulting company.[10]

[7] (Reader's Digest, 1977, p. 536)
[8] (Orwell, 1946, p. 252)
[9] (Sjodin, 2011, p. 3)
[10] https://en.wikipedia.org/wiki/Elevator_pitch

Whenever I'm invited to speak at a user group or corporate training session, one of the first questions I ask the event planner is, "In an ideal world, what can I do that would most benefit your audience?" I am always surprised by their answer. In most cases, they don't need something complex. Rather, it's something simple that they are puzzled about. There's an old medical saying: "If you hear hoofbeats, it's more likely to be horses than zebras."

In other words, start with the basics and then build from there. Bring your audience along with you. My wife was telling me of a common situation at the medical conferences she's attended. The presenter stands up and says, "They only gave me 15 minutes, but I need at least 90 minutes to explain this. So, here's a couple of slides; you can learn the rest from my handouts."

This is failure at multiple levels: insufficient time from the conference organizer vetting their speakers, the disdain of the presenter for their audience, and, worst, a lost opportunity to educate—at least a little—the audience in a technical subject that could spark their interest to learn more after the session. The best visuals in the world won't solve this problem, but a clearly thought-out elevator speech can help provide focus.

An elevator speech boils our message down to its core, the "one key thing" we want our audience to take away with them. Far too often, instead of condensing, we keep adding points, until we have a laundry list of the "absolutely, positively, most important 17 concepts you must remember from this speech." Sigh...that's a hopeless task for anyone.

"If people don't understand the change you want them to make, they can't make that change," writes business consultant Dianna Booher, whom we first met in Chapter 1. "If they don't understand your information or explanation, you will have less of a chance to change their minds about that issue. If instructions are too complex, people will likely resist the effort to follow them—or fail to accomplish the task. Simplicity and persuasion are intricately linked."[11]

The Four-Step Process of Persuasion

When it comes to persuasion, we can describe the process in four steps using the mnemonic AIDA:

- **Attention (A).** Hook their attention.
- **Interest (I).** Raise their interest.
- **Desire (D).** Make the benefits they get desirable.
- **Action (A).** Issue a clear Call to Action and tell them the next step.

[11] (Booher, 2015, p. 35)

According to Terry Sjodin, an elevator speech has several key characteristics:

- It is a clear, brief message with a purpose.
- Its sole function is to intrigue the listener to learn more.
- It has structure.
- It has a close.
- It helps you earn the right to be heard.[12]

Start your planning with an elevator pitch; fit it into two to three sentences. Use this to help you figure out what is most important to you. While you can cover multiple ideas in one message, there needs to be a single message that is most important. Don't spread the attention of your audience across multiple points. A knife cuts best when it has a sharp edge.

Focus, Then Focus Some More

E. St. Elmo Lewis, an American advertising advocate who was posthumously inducted into the Advertising Hall of Fame, foretold today's media landscape in 1903, when he wrote, "The mission of an advertisement is to *attract* a reader so that he will look at the advertisement and start to read it; then to *interest* him, so that he will continue to read it; then to *convince* him, so that when he has read it he will believe it. If an advertisement contains these three qualities of success, it is a successful advertisement."[13] Change the word "reader" to "viewer" and his advice still applies more than a century later.

We are not trying to impress viewers with what we know or how smart we are; we want to impress them with an urgency to *do* something.

Given the opportunity, we could easily list multiple "key points" we want to make in our message. But no viewer wants to watch a laundry list. She just wants to learn about what she is interested in. Dale Carnegie's "What's In It For Me?" (WII-FM) remains uppermost in her mind.

To help you figure out what's most relevant in your message, make a list of all the points you want to make, all your benefits, all the reasons why someone should pay attention to your message. Then, draw a line down the middle of a sheet of paper. On one side, write the positive aspects (pros) of your argument, on the other any potential negatives (cons). (Salespeople call this a "Ben Franklin close.") This allows you to determine possible objections from your audience then find positive statements to overcome them. Next, combine, consolidate, and compress all your points into three. Why three?

[12] (Sjodin, 2011, p. 12)
[13] (Lewis, 1903, p. 124)

Three points are easy for the audience to remember, with a powerful sense of rhythm that echoes across history:

- Faith, hope, charity
- Of the people, by the people, and for the people
- Beginning, middle, end
- Faster, higher, stronger
- Stop, look, and listen
- On your mark, get set, go

Your audience has a limited attention span. Present too many points and you'll lose them.[14] Presenting in groups of three forms a powerful rhythm that we can leverage to improve the visceral power of our presentation. Watch how often this rhythm of three occurs in the ads you watch or read.

How Many Benefits Are Enough?

Many times, we wrestle with how many benefits we should provide to entice the maximum number of people to respond to our offer. A seminal 2011 marketing study called, *The Presenter's Paradox* studied[15] whether offering more benefits made an offer more persuasive. What the authors discovered was that "when making an offer, communicators intuitively think more is better. ...But listeners don't look at the situation in the same way. Instead, they 'average' all the pieces of information they hear and walk away with a single impression. ...Not only did 'more' add less, it actually harmed the rest."[16] In other words, limiting what's contained in an offer makes it more powerful. They found the most successful number of benefits was about two.

Write Short

During my research, I enjoyed reading *How to Write Short* by Roy Peter Clark. So much of it pertains to what we are talking about. For example, "In the digital age, short writing is king. We need more good short writing—the kind that makes us stop, read, and think—in an accelerating world. A time-starved culture bloated with information hungers for the lean, clean, simple, and direct. Such is our appetite for short writing that not only do our *long* stories seem too long, but our *short* stories feel too long as well."[17]

Carl Hausman shares his three tips for better writing:

- Keep sentences short.
- Write in a conversational tone.
- Make very short sentences carry the weight.

"Of the people, by the people, and for the people" fascinates me. Not only is it a rhythmically powerful set of three phrases, each phrase itself is also a set of three. This repetition of a powerful rhythm is what helps make this so memorable.

When designing your message, remember the audience is always asking, "What's In It For Me?" (WII-FM).

[14] (Reader's Digest, 1977, p. 521)
[15] (Weaver, Garcia, & Schwartz, 2012, pp. 445–460)
[16] (Booher, 2015, p. 83)
[17] (Clark, 2013, p. 4)

This last is a potent technique and dirt simple. When you want to make an impact and get across a key point, make it a very short sentence.[18]

In these days of communication through tweets, "short" has developed an entirely new meaning. But, simply because a story is short does not mean it can't be compelling. Years ago, I read of Ernest Hemingway being challenged to write a story in less than ten words. He needed only six: "For Sale: Baby Shoes. Never worn." As the father of two and grandfather of two more, the sheer overwhelming emotion in that story continues to ricochet in my mind.

William Strunk, Jr., wrote *The Elements of Style* in 1918. Dr. Strunk was famous for his rules on writing style. For example, take Rule #13: "Omit needless words." When Dr. Strunk wrote "Omit needless words," he then added this paragraph:

> Vigorous writing is concise. A sentence should contain no unnecessary words, a paragraph no unnecessary sentences, for the same reason that a drawing should have no unnecessary lines and a machine no unnecessary parts. This requires not that the writer make all his sentences short, or that he avoid all detail and treat his subjects only in outline, but that he make every word tell.

Twitter has expanded their rules to allow 280 character comments. Perhaps, when you read this, even more. But, when it started it limited all comments to a 140 characters, forcing short, clear writing.

That paragraph contains 65 words, or 386 characters. Nice, but we live in the age of Twitter. What would this look like if it were reshaped for the limited characters of Twitter? Roy Peter Clark did the rewrite: "Write tight. A text needs no extra words as a drawing needs no extra lines. A sentence can be long with detail. But every word must tell." Wow! 137 characters, extracted from almost 400.

So, should you write long and cut back or write short to start with? As Clark summarizes, "There is no right answer, except for this: A good short writer must be a disciplined cutter, not just of clutter, but of language that would be useful if she had more space. How, what, and when to cut in the interest of brevity, focus, and precision must preoccupy the mind of every good short writer."[19]

For me, the most powerful way to write short for visuals is to combine the rhythm of human speech with solid content. Avoid sentences, use phrases.

Think Like a Poet

[18] (Hausman & Agency, 2017, p. 15)
[19] (Clark, 2013, pp. 118–122)

In poetry, words are precious and scarce. By this, I don't mean that every line needs to rhyme but that every word is important and needs to carry its full weight.

A good example is haiku. Traditional haiku uses three lines containing five, seven, and five syllables to reflect moments in nature. Originating in Japan in the mid-1600s, the "Great Four" haiku poets are Matsuo Bashō, Kobayashi Issa, Masaoka Shiki, and Yosa Buson. Their work, using a form set four centuries ago, is still the model for traditional haiku poetry today.[20] Here are three examples from the acknowledged masters of this craft:

An old silent pond...
A frog jumps into the pond,
splash! Silence again.

Matsuo Bashō

A world of dew,
And within every dewdrop
A world of struggle.

Kobayashi Issa

The light of a candle
Is transferred to another candle—
Spring twilight.

Yosa Buson[21]

Each of these poems contains less than 14 words yet still evokes a strong sense of place, pace, and emotion. Just as with haiku, images don't require lots of words to evoke powerful feelings.

The Battle Between "Think" and "Feel"

Jeremy Dean, in his article "The Battle Between Thoughts and Emotions in Persuasion" in PsyBlog, wrote this:

"Nowadays people tend to use 'I think' and 'I feel' interchangeably. For some this is a linguistic faux pas, but what about psychologically? Does it make any difference whether what you say is couched in 'thinking' or 'feeling' terms?

"...a new study published in the journal Personality and Social Psychology Bulletin finds this tiny difference can influence the power of a persuasive message.

"This suggests that if you want to persuade someone, then it's useful to know whether they are a thinker or a feeler and target your message accordingly. If you don't already know then the easiest way to find out is listen for whether they describe the world cognitively or affectively.

[The authors reported that] women were more persuaded by an ad for a new movie when it quoted reviews beginning with 'I feel.' Men, however, were more persuaded by the same basic ad when it quoted reviews beginning with 'I think.'"[22]

[20] https://examples.yourdictionary.com/examples-of-haiku-poems.html
[21] https://www.readpoetry.com/10-vivid-haikus-to-leave-you-breathless/
[22] (Dean, 2010)

Meet the "Words of Power"

I've written that we need to "write short," think of our writing as more like poetry than prose, and focus on works that "punch." But, are there "power words" we can use in our writing?

Yes! Paraphrasing George Orwell in *Animal Farm*, "All words are equal, but some words are more equal than others."

First, use active words and phrases. Avoid sentences that contain "be," "have been," or "was." As well, most times, the word "that" is not needed. Next, consider the five points Joseph M. Williams presents in his book *Style*:

- Delete words that mean little or nothing ("kind of," "really," actually," and "in order to").
- Delete words that repeat the meaning of other words ("various" and "sundry").
- Delete words implied by other words ("terrible tragedy").
- Replace a phrase with a word ("in the event that" becomes "if").
- Change negatives to affirmatives ("not include" becomes "omit").[23]

Mary Embree, in *The Author's Toolkit*, offers several more suggestions:

- Write from the heart.
- If you would write, read—especially books that touch your feelings.
- Keep your writing simple. Avoid talking "down."
- Limit the use of adjectives, adverbs, and qualifiers.
- When in doubt about something you wrote, take it out.
- Write in an active voice.
- Limit use of similes, metaphors, and analogies.
- Avoid clichés, um, like the plague.
- Avoid slang, unless it specifically applies to your audience. Avoid jargon for the same reason.[24]

One of the definitive studies of "power words" was conducted by Gregory Ciotti. He found five words that were simple, efficient, and powerful in speaking persuasively to an audience:

- **The viewer's name.** (Dale Carnegie spotted this as well almost 90 years ago.) We all like seeing our name in print.
- **Free.** This is so powerful people will lean in its favor even when it comes with a hook.

[23] (Clark, 2013, p. 124)
[24] (Embree, 2010, pp. 33–45)

- **Because.** Asking a favor is fine, but asking a favor followed by a reason is much more effective.

- **Instantly.** This, along with "immediately" and "fast," are quick triggers for motivating human behavior.

- **New.** This is tricky; we want to buy new products, but from recognized brands. "New" is risky; find ways to minimize the risk. Stress the reliability of your company.

Gregory Ciotti ends his blog with this warning: "I can't stress enough—just as in the application of writing headlines that work—you must understand why these words are persuasive, and you must use them in the contexts that make sense for your audience and your business. If you just start slapping them on every piece of content you create for no apparent reason, you'll quickly see just how unpersuasive they can be."[25]

Every story is different, and styles that work for one won't work in the other. Still, there's no reason to set yourself up to miscommunicate. And, by the way, the word "please" is pretty darn magical, too. There's never harm in being polite.

To emphasize the importance of each word, **TABLE 3.1** shows some typical word counts for different presentation formats.

TABLE 3.1 Typical Word Counts for Different Presentations	
PRESENTATION	AVERAGE WORD COUNT
Meme	Less than 10
PowerPoint slide (optimal)	Less than 20
15-second video	Less than 30
Printed poster	Less than 50
30-second video	Less than 75

See What's in Front of You!

After you've designed your message and written your story, there is one last *critical* task: Proofread your work! The hardest challenge my students have is seeing what's *actually* on the screen. Instead, like most of us, they see what they *expect* to see.

[25] Gregory Ciotti, (December 6, 2012), The 5 Most Persuasive Words in the English Language, Copyblogger Media LLC,

Proofreading is not easy. For example, look at **FIGURE 3.4** again. It looks perfectly fine, except...this is supposed to be an ad for swimming *goggles*.

I can't teach you how to see. But I can personally testify that the darnedest mistakes crop up at the most inopportune times. What I found works for me when proofreading my work is looking at the same thing two different ways— say on-screen then on paper. Or on-screen, but using two different programs so the layout changes.

Anything to help me see the same text displayed differently. It always amazes me how many errors I catch that way, even when I think it's perfect.

KEY POINTS

Here's what I want you to remember from this chapter:

- Stories are deeply woven into the way we think and feel.
- Unless guided otherwise, our brains will invent a story for an image.
- In any story, change is essential.
- A workflow makes us more efficient.
- A message is the content, while a story is how that content is delivered.
- Focus your message, and target it at your audience.
- Write short, use active words; think poetry, not essays.
- Proofread your work!

PRACTICE PERSUASION

Rewrite the following 119-word paragraph for brevity. Specifically:

- For a 30-second movie trailer
- For an original 140-character Twitter post
- For a 15-word motion graphic video

A Tale of Two Cities (Charles Dickens)

It was the best of times, it was the worst of times, it was the age of wisdom, it was the age of foolishness, it was the epoch of belief, it was the epoch of incredulity, it was the season of Light, it was the season of Darkness, it was the spring of hope, it was the winter of despair, we had everything before us, we had nothing before us, we were all going direct to Heaven, we were all going direct the other way—in short, the period was so far like the present period, that some of its noisiest authorities insisted on its being received, for good or for evil, in the superlative degree of comparison only.

PERSUASION P-O-V

FRESH FISH

Thinking about writing short reminds me of the story quoted in The Art of Persuasion, called "Fresh Fish."

The fishmonger Klotnick has acquired a nice piece of wood to use as a sign, but he isn't sure it is big enough for what he wants to say: *Fresh Fish Sold Here Daily*. Just then his friend Rubinstein comes along, and Klotnick asks for his advice. Well, says Rubinstein, you don't want to say *Fresh*—that'll set people wondering if the fish maybe aren't so fresh. And you don't need *Sold*—would you give them away? *Here* is a mistake—where else would you sell them? And *Daily* you don't need—if the shop's open, you're selling. Come to think of it, the people can see the fish in the window—so why bother with *Fish*?[26]

[26] (Bridges & Rickenbacker, 1992, p. 112)

There's fashion in typefaces, just as there is in clothes. What is a type style but a suit of clothes for the letters of the alphabet? You can send your words out into the world looking as though they are in touch with modern trends, or stuck in a 1980s timewarp of big-haired power dressing; wearing an old tweed jacket with leather patches on the elbows, or as a total fashion victim.

SIMON LOXLEY[1]
BRITISH GRAPHIC DESIGNER AND AUTHOR

CHAPTER 4
PERSUASIVE FONTS

CHAPTER GOALS

The goals of this chapter are to:

- Explain what fonts and typefaces are
- Describe different font families and the emotions they suggest
- Provide tips on how best to use fonts in your projects
- Illustrate how to pick the right font for your message

THE YEAR WAS 1989; I was working as a marketing manager for Bitstream in Cambridge, Massachusetts. Bitstream was the first major digital type foundry to devote all its resources to typographic design and digital technology. My job was to launch a new series of typefaces specifically designed for Macintosh computers.

The right typeface ensures that your message gets read and remembered. And adds visual power to your printed words.

The company was founded in 1981 by legendary type designer Matthew Carter, along with Cherie Cone, Michael Parker, and others. A type foundry designs typefaces. At the time, we were competing with Adobe, Linotype, URW, and many others in the exploding world of fonts.

In those days, the world was making the transition from low-resolution computer screens and dot-matrix printers into the scalable world of laser printers and digital printing presses from companies like Heidelberg. When I joined Bitstream, I knew a lot about marketing but nothing about fonts.

As I walked into the basement of the three-story brick building that housed Bitstream, I walked into another world. In that vast, dimly lit space stood row after row of giant computers, each as tall as a man, clad in deep-blue sheet metal with a glowing green screen about the size of a legal sheet of paper. At each system was an artist, each of whom could draw beautiful high-quality letterforms freehand, designing new typefaces with the aid of these computers.

Sitting in meetings, discussing which fonts we should bring to the Mac market, listening to Matthew and other designers discuss the influence of different typefaces—was an unforgettable education for me. I fell in love with type, not as a designer but as a consumer.

About that same time, Kelsey Selander joined the company as vice president of marketing. A bundle of energy and enthusiasm, she also looked at type as a consumer. Her big insight, from my perspective, was that type needed to be marketed not on how the individual letterforms looked but based on the emotions they sparked in the viewer.

"The right typeface ensures that your message gets read and remembered," Kelsey wrote in a 1989 Bitstream type marketing brochure. "It enlivens and clarifies the meaning of your ideas. And adds visual power to your printed words."[2]

[2] (Bitstream Typeface Marketing booklet, 1989)

Designers have always known of the emotional power of type, as **FIGURE 4.1** illustrates. Kelsey's insight was to focus on the end user, the consumer of type. She kept asking, "How does this font make you feel?" She realized that the more clients knew about type, the more fonts they would use or request of their graphic artists.

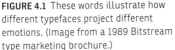

FIGURE 4.1 These words illustrate how different typefaces project different emotions. (Image from a 1989 Bitstream type marketing brochure.)

FIGURE 4.2 Printing books in England during the Renaissance: sorting letters on the left, typesetting in the middle, and printing on the right. (Image from the Walker Art Library / Alamy Images.)

"Typography is just that: idealized writing," poet and typographer Robert Bringhurst wrote. "In a badly designed book, the letters mill and stand like starving horses.... In a well-made book, where designer, compositor, and printer have all done their jobs, ...the letters are alive. They dance in their seats."[3]

A Quick History of Type

Most of us are familiar with the name of Johannes Gutenberg, who is credited, in the West, as the inventor of the printing press, as illustrated in **FIGURE 4.2.** It took him about 20 years to develop all the elements necessary for a working press, which was completed in 1439.

Describing this process, British graphic designer and author Simon Loxley wrote, "When people refer to the birth of printing in Europe, what they actually mean is the birth of type. Movable type—individual letters that could be arranged, edited, printed from, then dismantled and reassembled to print again in a new configuration—this was the real breakthrough, the spark that fired the printing revolution that was to sweep through Europe during the rest of the century.[4]

[3] (Bringhurst, 2013, p. 19)
[4] (Loxley, 2004, p. 7)

FIGURE 4.3 Gutenberg's masterstroke was his invention of movable type. (Image Credit: Gabe Palmer / Alamy Images.)

Gutenberg's many contributions to printing include:

- Mechanical movable type (**FIGURE 4.3**)
- The invention of a process for mass-producing movable type
- Adjustable molds for casting type
- The use of permanent oil-based ink for printing books
- The use of a wooden printing press similar to the agricultural screw presses of the period[5]

The introduction of the printing press, along with the mass production of paper to go on it, revolutionized bookmaking. The wide availability of books massively altered the structure of society. Texts no longer belonged principally to the church. People, not just clergy, demanded to learn to read, creating a need for education. Ideas spreading across borders threatened established power structures. Greater education created a middle class. The rise of the middle class fueled the Renaissance. Without question, today's information-based economy is a direct result of Gutenberg's invention.

Type designer Matthew Carter once wrote, "Type is a beautiful group of letters, not a group of beautiful letters." How letterforms work together is critical to readability.

The Design of Type

The creative process of designing typefaces is called "type design." (**FIGURE 4.4** shows some of the complexity of this.) Type designers work in type "foundries," so called because, initially, movable type was created by pouring molten lead into molds. When it comes to digital fonts, designers are sometimes called "font developers" or "font designers."

These days, "fonts" and "typefaces" have essentially become synonyms, but there are differences. Here are some key terms about type:

- **Type.** Everything related to typefaces and their design.
- **Typeface.** A family of fonts of a similar design. *Helvetica* is a typeface. We often shorten this to "face."
- **Font.** A specific design of a typeface. *Helvetica Italic* is a font.
- **Weight.** Condensed, bold, and italic are all weights of a typeface; also describes the thickness of a glyph.

[5] https://en.wikipedia.org/wiki/Johannes_Gutenberg

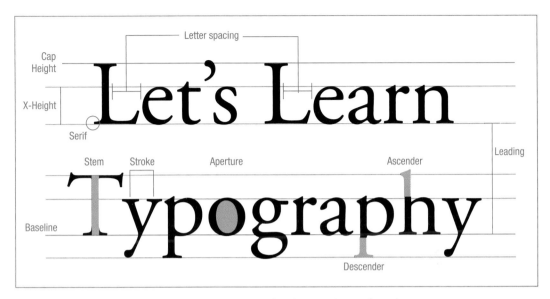

FIGURE 4.4 This illustrates the complexities designers consider when creating typefaces that the rest of us take for granted.[6]

- **Glyph.** The shape of the letters in a typeface. "3" is a number. The shape of the "3" is the glyph. "Letterform" is a synonym for glyph.
- **Typography.** A way of setting and arranging type to make it easy to read, without losing the emotions inherent in the font.
- **Line spacing.** The vertical spacing between lines of text.
- **Tracking.** The horizontal spacing between glyphs; also called "letter spacing."
- **Kerning.** Manually adjusting the spacing between selected pairs of letters.

Even with these definitions, "typefaces" and "fonts" are often used interchangeably, both in this book and in real life.

As an interesting piece of trivia, typeface designs in the United States cannot generally be copyrighted, while the names of the designs can. This is why the same design—Helvetica, for example—has so many different names. Helvetica, Swiss, Universe, Arial, Triumvirate, Pragmatics, and Nimbus Sans are all different names for the same font design from different type houses.

[6] https://dtc-wsuv.org/
jrushing17/dtc478/final

There are close to 100,000 fonts in the world.[7] The Latin alphabet contains more than 600 characters,[8] while more fonts and characters are created every day (all before we add emojis into the mix!). Thankfully, however, we don't need to learn to use each letter individually. Instead, there are six main "families" of fonts in the Latin alphabet, as opposed to the different word shapes in languages such as Chinese, Arabic, or Hebrew:

- Serif
- Sans serif
- Script
- Blackletter
- Monospace
- Specialty

Each of these families has many typefaces within it, each with a particular role to play in connecting your message to specific emotions in your audience.

Compared to print, video and the web are low-resolution. A typical image for a half-page magazine ad is about 20 megapixels. A high-definition (HD) video image is only 2.1 megapixels. (Old-fashioned standard-definition (SD) media was far less resolved at only 0.3 megapixels!) Even a 4K video image is less than 9 megapixels. There isn't a lot of pixel resolution to work with in digital media.

Additionally, text in moving media needs to be read quickly. Typical television ads hold text on-screen for only two to four seconds. Readability is critical. This means that font selections must be focused on clarity, so the eye can read these messages quickly.

We use fonts to convey a message and wrap it in an emotion. However, if we are not careful, there can be a disconnect between the message we send and the fonts we use.

FIGURE 4.5 perfectly illustrates when the emotions in a font conflict or complement the message being sent. Compare the spun-sugar feeling of Edwardian Script with the solidity of Giza. You can *feel* those Giza pipes are solid cast iron.

> If typefaces and their design interest you, find a copy of *The Elements of Typographic Style* by Robert Bringhurst. Robert has the eye of a designer and the soul of a poet. Hermann Zapf described this book as "The Typographer's Bible." It makes for fascinating reading.

FIGURE 4.5 The spidery and ornamental shapes of Edwardian Script (top) are totally at odds with the message, while the slab-serif, solid lettering of Giza (bottom) tells the viewer that these pipes would never dare to leak. Ever.

[7] (Garfield, 2011)
[8] (Bringhurst, 2013, p. 303)

We have two goals when choosing fonts for our images. First, and most important, we want fonts that are clear and easy to read. Second, we want to use fonts that emotionally reinforce our message, as illustrated in our plumbing logo in Figure 4.5.

There are no "wrong" fonts, but, given a bit of thought, you'll discover that some fonts fit your message better than others. Remember those six type families mentioned earlier? They can help you find the right font for your project.

Serif Fonts: The Voice of Tradition

"Serif" comes from the Dutch meaning "line, or stroke of the pen." I think of serifs as fonts with little "feet" at the edges of a glyph. Look at the shapes of the letters in **FIGURE 4.6**. See the little feet at the bottom of the "A" or the "jaws" at the ends of lines in the letters "C" and "E." Those are serifs. (Um, no, designers don't call them "feet" or "jaws," but they should.)

Things seem easier to do when they are easier to read.

Serifs are modeled after the way the Romans carved letters into clay. Most often, the edges of a letter had a serif, formed by the tool they used to form the letters.

Serif glyphs descended from Roman lettering of more than 2,000 years ago. In Rome, there's a statue to the Emperor Trajan sitting at the top of a tall, marble column. At the base of the column is lettering explaining why Trajan was being honored. These letters, hand-carved in the stone, are considered so clear, so readable, and so beautiful that they have been studied and copied for centuries. The Trajan font, **FIGURE 4.7,** was modeled after those letters.

AbCdEfg

FIGURE 4.6 A typical serif font; ITC Galliard, in this case.

ABCDEFG

FIGURE 4.7 Trajan, a font designed in 1989, is based on stone carvings from ancient Rome.

You'll notice that I don't mention Times New Roman. The reason, aside from the fact that it is way overused, is that the glyphs are small and hard to read on-screen. There are many serif font options that provide better readability for digital delivery.

This long history is a key personality attribute of serif typefaces. They evoke feelings of tradition, history, reassurance, and professionalism. They are also useful when talking about money, because serifs conjure feelings of banks and conservative financial institutions.

Take a look at the five fonts in **FIGURE 4.8**. You've seen one of the first three in just about every book you've ever read. Notice that Baskerville and Palatino have thicker glyphs. This allows them to look good on lower-resolution display screens, though Palatino takes more horizontal space than Baskerville. Garamond is a lovely, refined font, though not as dark as Baskerville or Palatino.

Bodoni and Onyx scream "high-fashion." They are tall and thin, with delicate strokes and crossbars. However, these thin lines raise a big caution. Fonts like Bodoni and Onyx have such thin serifs that they disappear when displayed on low-resolution devices like monitors at small point sizes. This

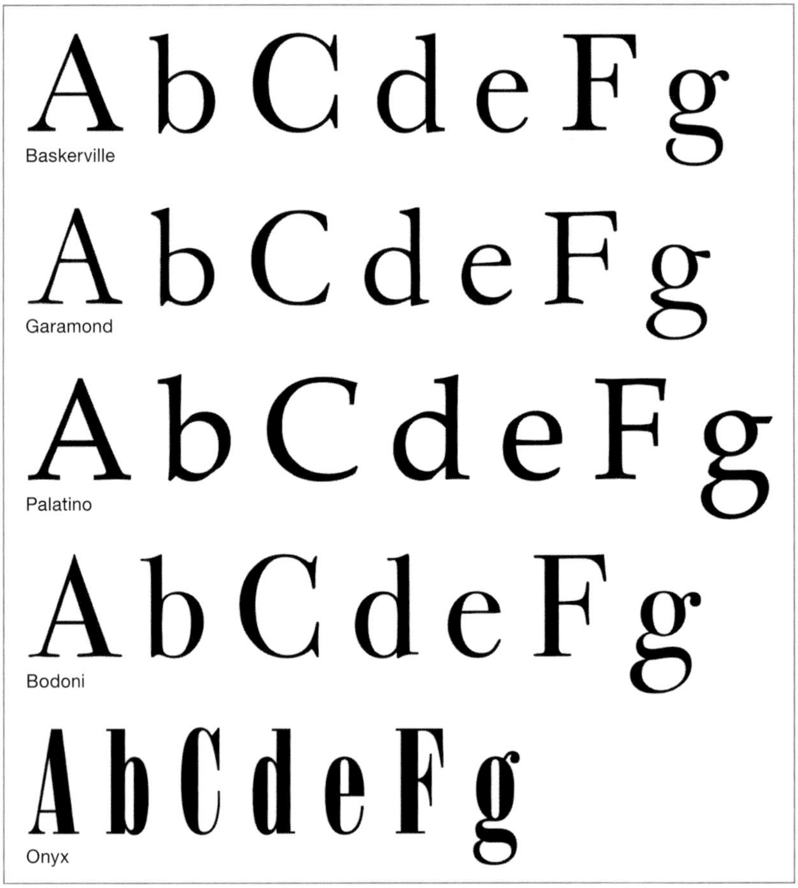

FIGURE 4.8 Examples of popular serif fonts. These all use the same point size, but I increased the letter spacing to make it easier to see the form of each glyph.

Baskerville

Garamond

Palatino

Bodoni

Onyx

often converts an "e" to a "c" or an "o" to a "u" because parts of the letterform get lost. It is important to keep in mind that not all serif fonts look good on digital displays.

When looking at font design, a good place to start is the lowercase "e" and "g"—"e" because that's where too-thin lines are most visible and "g" because that's where designers give their art a freer rein. In the fonts in Figure 4.8, see how little space there is above the bar of the "e" in Baskerville or Garamond compared to Palatino. Then compare the Onyx "g" to Garamond. Glyphs are an art in and of themselves.

Serifs typefaces project feelings of:

- Tradition
- Respect
- Reliability
- Comfort

Serif italic fonts (**FIGURE 4.9**) extend the creative fancy of the type designer into new areas. Unlike any other type family, serif italics differ wildly from their non-italic version. I first met Galliard, which was designed by Matthew Carter, when I worked at Bitstream. The hand-drawn spirit of its italic letters continues to fascinate me. As does the swooping elegance of Garamond italic, or, the "let's keep things moving" nature of Baskerville italic.

We use fonts to convey a message and wrap it in an emotion.

Baskerville

Garamond

Palatino

ITC Galliard

FIGURE 4.9 Unlike any other font family, serifs have a radically different look when used in italic form.

Spend time exploring the look of different fonts—enlarge them to 400 points and just admire the art. When it comes to emotions, italic serif fonts reflect:

- Style
- Fashion
- Grace
- Elegance

Since serif fonts represent traditional values, what fonts point us into the future? That is the role of sans serif type.

Sans Serif: The Voice of the Future

Sans serif, which is French for "without serifs," are fonts that don't have time for the niceties; they have business to conduct. Gone are all those little feet.

Sans serif fonts (**FIGURE 4.10**) are designed with today's digital media in mind: solid glyphs with not a lot of variation in the stroke width. These fonts are designed first for readability, second for art.

Futura is a popular workhorse, as is Century Gothic (though Century Gothic has more grace in its design, like putting a font in a business suit). I use Impact whenever it is *essential* you pay attention to a title, though it is a terrible body copy font. I'm a new fan of Calibri. It doesn't take a lot of space, looks very clean on the screen, and is eminently readable. Then, compare each of these to the soft flowing letters of Optima. Like a serif, the width of the strokes varies within the letter, though it remains solidly a sans serif font. It is an elegant, yet still modern, font.

What Happened to Helvetica?

You'll notice Helvetica is not in this list, for the same reason that I omitted Times New Roman—Helvetica is *waaaay* overused. It is a lovely font but is used everywhere. Just because it is the default for most computer applications does not mean it is the best choice for you. Helvetica needs a rest. As typographer Erik Speikermann wrote, "Most people who use Helvetica use it because it's ubiquitous. It's like going to McDonald's instead of thinking about food. Because it's there, it's on every street corner. So let's eat crap, because it's on the corner."[9]

I don't think Helvetica is "crap"; it is a lovely typeface that is far too popular. I should also mention that Helvetica is the only typeface to have inspired a feature film: *Helvetica*, directed by Gary Hustwit and released in 2007.

[9] Erik Spiekermann, Typographer. https://typography.guru/quote/

FIGURE 4.10 These popular sans serif fonts all share the same point size and letter spacing. Unlike serifs, there are no "feet." All the strokes in the same font are about the same thickness.

AbCdeFg
Futura Medium

AbCdeFg
Century Gothic

AbCdeFg
Impact

AbCdeFg
Calibri

AbCdeFg
Optima

Sans serif typefaces project:

- Stability
- Strength
- Objectivity
- Cleanliness
- Modernity
- Progressiveness

When in doubt, sans serifs should be your first choice for online and media-related messages. Highly readable, strong, and clear, these may not have the evocative impact of serif fonts, but they make sure your message gets across.

FIGURE 4.11 illustrates a "waterfall," the same font displayed at different point sizes. Note how much darker Futura is compared to the same point size of Palatino, and Palatino is darker than many other serif fonts. Darker typefaces tend to be easier to read, especially if the background is busy. Also notice how sans serif fonts are much easier to read at smaller point sizes.

Since we are talking fonts, this book uses Meta Serif Pro Book for body text, Meta Pro Bold and Black for titles, and Ronnia Light for side-bars, captions, and other incidental text. Oh! And Meta Pro Normal for the P-O-V stories at the end of each chapter.

> Serif: Palatino
>
> 15-point: The quick brown fox jumped over the lazy dog.
>
> 12-point: The quick brown fox jumped over the lazy dog.
>
> 10-point: The quick brown fox jumped over the lazy dog.
>
> 7-point: The quick brown fox jumped over the lazy dog.
>
> Sans Serif: Futura Medium
>
> 15-point: The quick brown fox jumped over the lazy dog.
>
> 12-point: The quick brown fox jumped over the lazy dog.
>
> 10-point: The quick brown fox jumped over the lazy dog.
>
> 7-point: The quick brown fox jumped over the lazy dog.

FIGURE 4.11 See how sans serif fonts tend to be more readable at smaller point sizes?

Script Fonts: Handwriting for Computers

Not all of us have perfect handwriting, I certainly don't. Script fonts (**FIGURE 4.12**) were invented to reflect the personal nature of handwriting, but in a form that computers can easily display.

Script fonts, even more than serif italic, reach new heights of artistic fancy. Whether it's the yeoman clarity of Brush Script, the teacher-at-the-chalkboard look of Chalkduster, the careful hand-printing of Noteworthy, or the extravagant elegance of Zapfino, script fonts are just plain fun.

Unlike serif and sans serif, script faces generally have only one weight, meaning, for instance, that there would not be an italic version.

But, because of that fun, we sacrifice readability. It takes at least twice as long to read a script font as it does serif or sans serif. It also means that the reader needs to *want* to read your text. Often, they see script text and say, "This is too hard to read; I'll read it later." Only they never come back.

Use script fonts when you know the reader is interested in your subject and when they have time to parse the text. Be sure to use script faces in larger point sizes; *never* use script as a body text.

Script fonts project:

- Elegance
- Formality
- Affection
- Creativity
- Personality

Blackletter: Extreme Script

Blackletter fonts (**FIGURE 4.13**) take us back to medieval times—or a heavy-metal concert. They conjure Gregorian chants and times of dragons, fair maidens, and quests. But they are just *awful* for readability. Feel free to use these for logos, but if you expect anyone to actually *read* these, allow an extra week.

You'll also see blackletter fonts used specifically because they are emotionally so different from the message. Heavy-metal and grunge rock groups use them to emphasize the irony.

Blackletter fonts project:

- Antiquity
- Religion/spirituality
- Classic
- History

Cloister Black - BT

Blackmoor - LET

Monospace: Return of the Typewriter

You may have heard of typewriters. (Think of them as "manual printers.") Those machines all used monospace fonts (**FIGURE 4.14**). A monospace font is one where all the letters are the same width.

I'm old enough to have actually *used* a typewriter for all my school reports. Typewriters were magical to me because they meant I no longer needed to handwrite everything. As fast as I could think, I could put words on the page with perfect legibility except, ah, deleting a typo required either retyping the entire page or smearing the paper with white glue. Typewriters were better than handwriting, and much more legible, but I'm far happier using a word processor today.

The 800-pound gorilla of monospace fonts is Courier. It was the most popular font on the dominant typewriter of the day: the IBM Selectric. (Sigh... Hands-down, the best keyboard ever invented.) In Figure 4.14, look at the clean lines, even strokes, and sheer readability of Courier. Then, look how American Typewriter adds some style quirks. Letter Gothic was designed for speech-writing, while Andale Mono is a blend of Courier's clean lines with American Typewriter's style.

Every character in a monospace font has the same width. The reason for this was that a typewriter only moved the paper in front of the keys a fixed distance horizontally for each letter. Thus, the glyphs themselves all had to have the same width. As you can imagine, this caused problems when trying to make an "i" and an "m" look comfortable together.

Variable-space fonts (**FIGURE 4.15**) give different widths to different letters; for example, the "l" and "i" are narrow, compared to a "w" or "m." Typewriters, however, couldn't deal with letters of different width, so each letter was designed to take the same amount of horizontal space.

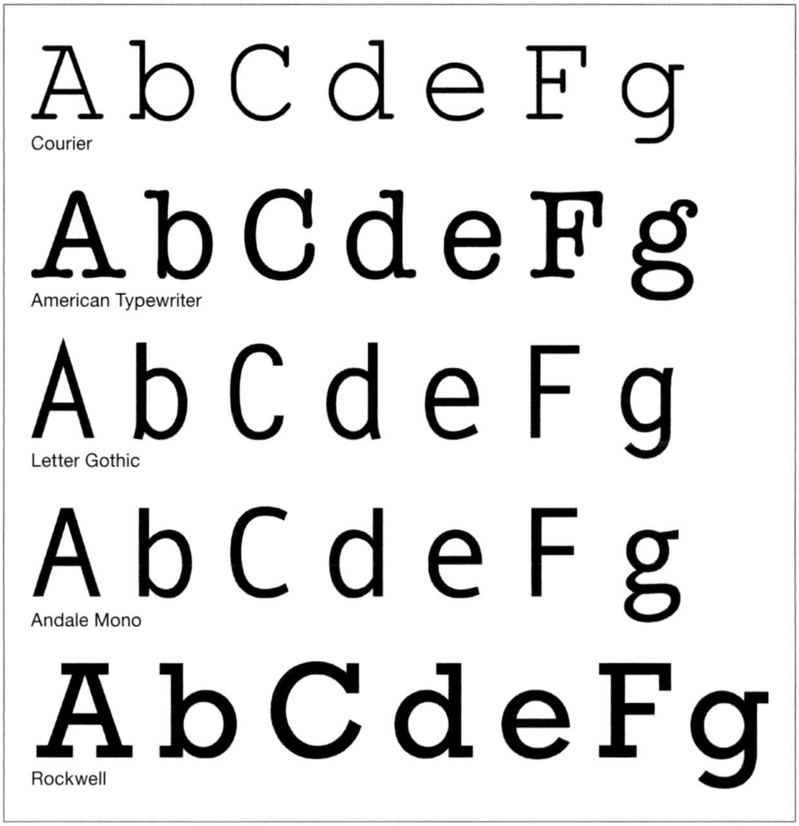

AbCdeFg

Courier

AbCdeFg

American Typewriter

AbCdeFg

Letter Gothic

AbCdeFg

Andale Mono

AbCdeFg

Rockwell

FIGURE 4.14 The top four are actual monospace fonts. Rockwell is a monospace "ringer," a variable-spaced font impersonating monospace. Why? To save space on the page.

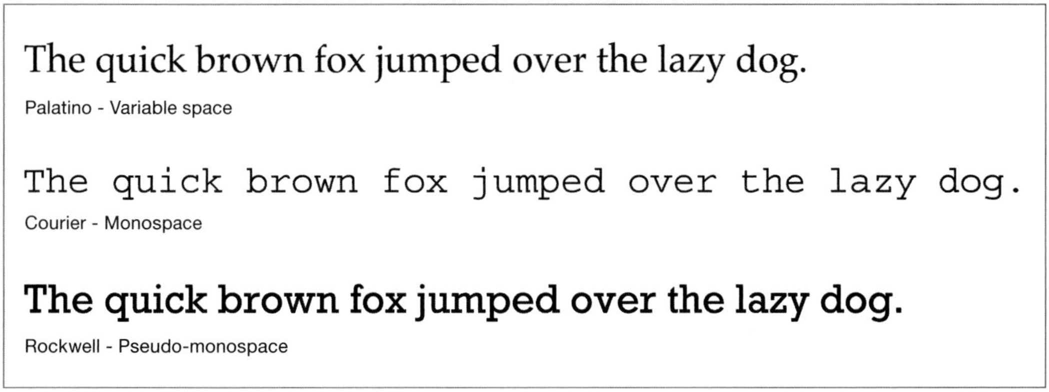

The quick brown fox jumped over the lazy dog.

Palatino - Variable space

The quick brown fox jumped over the lazy dog.

Courier - Monospace

The quick brown fox jumped over the lazy dog.

Rockwell - Pseudo-monospace

FIGURE 4.15 Compare the glyph widths in a variable-space font (top), a monospace font (middle), and a "pseudo-monospace" font (bottom).

Monospace fonts provide a historical feeling. Telegrams, radio bulletins, and scripts all look better in a monospace font. Portable printers before computers (we called them "typewriters") all used monospace fonts.

Monospace fonts project:

- A specific period, such as the 1940s and 1950s
- Bulletins and fast-breaking news
- Media, such as television and film scripts
- Antique equipment and communication

Specialty—Creativity Runs Amok

If serif fonts are wallflowers, stoically doing their job without calling attention to themselves, specialty fonts (**FIGURE 4.16**) are dancing on the table wearing only a wide smile and scarf, holding a flower between their teeth. Designed strictly for titles, their sole goal is attracting attention—and the more attention, the better.

FIGURE 4.16 Specialty fonts are narcissists, constantly calling attention to themselves. They are designed solely for titles and sheer exuberance.

These are best used for titles, as they wear out their welcome quickly. Many don't even have lowercase letters; they aren't that modest. *Everything* with them is a CAPITAL LETTER!

Be careful with fonts like Desdemona and Jazz, among others. You need to keep the point sizes large enough that the thin lines in the letterforms don't break up.

Specialty fonts project, well, just about anything, as long as you keep looking at them. They are very narcissistic; see **FIGURE 4.17**.

FIGURE 4.17 Rosewood was feeling left out and asked to be included.

Centuries Before Emojis Came Dingbats

There's also a class of "letters" called "dingbats." Dingbats were the emojis of the Renaissance. These are printer's ornaments that are nothing but symbols; they are like emoji, but centuries older. The most popular dingbat family is called Wingdings, illustrated here.

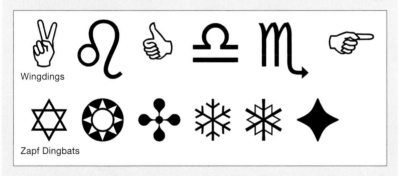

Wingdings

Zapf Dingbats

Font Techniques

Never forget that our number one goal whenever we add text to an image is readability. Why? Because we can't deliver our message if the viewer is unable to read it. Everything takes second place to catching and holding the eye of the viewer long enough to tell our story.

FIGURE 4.18 An example of fonts in service to readability. (Photo courtesy of Philip Snyder.)

Thinking of fonts and readability, Philip Snyder sent me this photo (**FIGURE 4.18**) of a Geico banner, towed by a small plane over the beaches of Long Island each summer. Designed to be read from a distance of 1,000 feet, look at the simplicity of the design, the use of sans serif fonts, the large text, and a close-up of the spokes-lizard.

With readability as our goal, let's take a look at a variety of techniques that will help us improve the readability of our text.

Avoid ALL CAPS

Don't put text in all caps if you want anyone to actually read it (**FIGURE 4.19**). For that matter, don't put body text in a script face. Actually, this whole paragraph is completely unreadable—a classic example of incomprehensibility using perfectly clear fonts.

FIGURE 4.19 Put text in all caps when you want to be sure no one reads it.

> *ONE OTHER NOTE:*
> *AVOID PUTTING BLOCKS OF TEXT*
> *IN ALL CAPS UNLESS YOU INTEND*
> *THAT NO ONE IN YOUR AUDIENCE*
> *SHOULD READ IT.*

Use Drop Shadows

Many times, we need to put a title into brighter areas of the image, such as white text against the sky. Or, in the example in **FIGURE 4.20**, the text is the same color as the background. In either case, the text becomes unreadable. In these situations, putting a drop shadow under the text instantly improves readability.

FIGURE 4.20 Text the same color as the background was added to this image. It is impossible to read. (Image Credit: pexels.com)

FIGURE 4.21 Simply adding a drop shadow to the text makes all the difference. (Image Credit: pexels.com)

The difference in readability between Figure 4.20 and **FIGURE 4.21** is obvious, yet all I did was add a drop shadow to the text. The color of the text is the same in both images. Regardless of whether you are creating still or moving images, *always* add drop shadows to your text, *except* under two conditions:

- If the text color is black, drop shadows will only make the text harder to read.

- If the background is black, drop shadows will disappear.

Recommended Drop Shadow Settings

Here are my recommended drop shadow settings (these should work in virtually every application):

- **Angle.** Adjust shadows to fall down to the right (Photoshop: 135°).

- **Opacity.** 95%.

- **Distance.** Between 5–15 pixels.

- **Spread.** 5–10 pixels.

Kern Title Text

Kerning adjusts the horizontal spacing between two letters. I first ran into the importance of kerning years ago reading about the challenges in Biblical translation where all source text was handwritten. Scholars pointed out how difficult it is to precisely determine meaning because of irregular letter spacing, as shown in **FIGURE 4.22.**

Every glyph in every typeface on all computers, in addition to its shape, includes spacing information, called its "bounding box," that describes how

nowhere

FIGURE 4.22 Kerning helps clarify meaning. Is this text saying "no where" or "now here"? Without better kerning, this could go either way.

FIGURE 4.23 Kerning adjusts the horizontal spacing between two characters. The boxes on the left indicate bounding boxes for each letter.

wide and tall each character is, as well as how it fits with the character next to it. (I expanded the spacing between the bounding boxes in the text on the left of **FIGURE 4.23** to make the boxes easier to illustrate.)

But, as the text in Figure 4.23 illustrates, some letter combinations create wide spaces between letters. We fix that with kerning. In general, you *don't* kern body text, but *do* kern title text.

Kerning pairs of characters manually allows us to create a much more pleasing look.

Kerning in Photoshop

In Photoshop on a Mac, place the text cursor between the two letters you want to adjust and press Option+left/right arrow. (On Windows, press Alt+left/right arrow.) This moves the two letters closer together or farther apart.

Tighten Line Spacing for Titles

Another area where tweaking makes a big difference is line spacing titles. Line spacing determines the vertical distance between two lines of text. The default values are set by those bounding boxes I mentioned earlier, and, by default, they expand vertical line spacing by 20 percent to make body text easier to read. The problem is that this makes titles too open, as you can see from the left title text in **FIGURE 4.24.**

FIGURE 4.24 Line spacing adjusts vertical position between lines. Reduce line spacing 20 percent to 30 percent for almost all titles.

Movie Movie Title Title

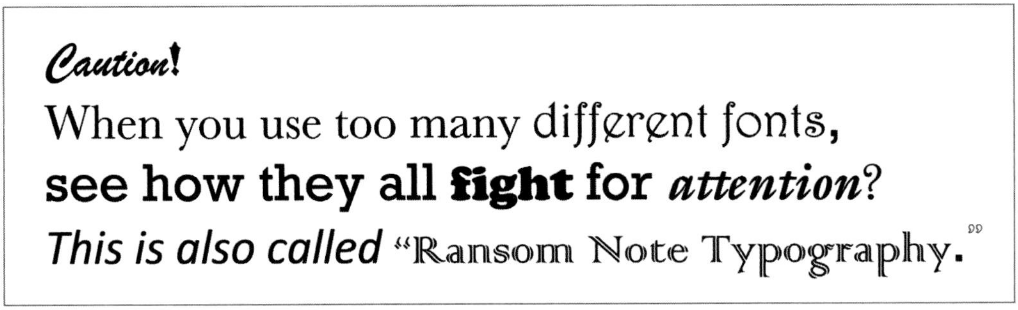

Caution!

When you use too many different fonts,

see how they all **fight** for *attention*?

This is also called "Ransom Note Typography."

FIGURE 4.25 Need I say anything more? There are 11 different fonts in this visual catastrophe.

Reducing the line spacing for titles 20 percent to 30 percent smaller than the point size of the font will make your titles look more readable and professional. As an example, look how much "punchier" the title on the right looks, compared to the spaced-out version on the left. The title on the right looks solid, forcefully highlighting your message.

Avoid Ransom Note Typography

When in doubt in any situation calling for text, just use one or two fonts: one for titles and other for body text. The days of impressing your audience with multiple typefaces are long gone (**FIGURE 4.25**). Today, folks just want to read what you wrote.

Avoid making stupid mistakes with your type. To get an idea of the potential icebergs lurking in your path, do a Google search for "Examples of bad typography." By understanding the differences fonts make in telling your message, you can avoid these obvious mistakes.

Choosing the Right Fonts

When choosing fonts, remember that digital media tends to have lower resolution than print. So, use fonts that:

- Are different from the default fonts of your system (give Helvetica and New Times Roman a rest).
- Are easily readable, especially when used at smaller point sizes.
- Reflect the style and emotion of your message, especially for titles and full-screen graphics.
- Use relatively even strokes in the glyph; avoid glyphs with very thin strokes.
- Contrast with the background color and pattern.
- Finally, give your audience enough time to actually *read* them.

KEY POINTS

Here's what I want you to remember from this chapter:

- The fonts we use and how we use them trigger emotional responses in our audience before they even read a word of the text.
- Pick fonts that represent the style and emotional tenor of your message.
- Pick fonts for their readability first, then their design.
- Sans serif fonts tend to read better on digital media.
- Of the two principle font families: serif is traditional; sans serif is modern.
- Fonts with thin elements don't look good on low-resolution screens.
- Always add drop shadows to title text, except in limited situations.
- Always kern and letter space title text.

PRACTICE PERSUASION

Using a word processor, set this headline in different fonts and font sizes. Imagine you are creating two versions: one for a plumbing hardware store and the other for a cupcake boutique.

How does the choice of font affect the emotion and urgency of the message?

New Look! New Colors! New Style!

PERSUASION P-O-V

A PRESS RELEASE THAT SAYS NOTHING

In the chapters on writing and fonts, I've emphasized clarity, readability, and thoughtful consideration of your audience and message. Here's the exact opposite: an actual press release so opaque and self-absorbed as to be completely unintelligible. (I deleted the name of the specific company and product for all the obvious reasons.)

This was sent to a general technology press list, in hopes of generating press coverage. I read this a dozen times and still have no idea what they are talking about or why I should care. This illustrates what happens when what you know gets in the way of clearly explaining it.

"1st Oct 2019: The global product design, development, and IoT company, ___, who is the development and design service provider behind the ___ Smart Ring product has just released a Smart Ring white paper to help brands, manufacturers, and service providers understand this emerging wearable market, and the technology, and gives guidance to create a successful Smart Ring strategy.

"Over the past five years, the Smart Ring market has grown steadily—primarily driven by tech startups. ___, the global product development and IoT firm, and developer of several signature Smart Ring products, predicts that Amazon's Echo Loop Smart Ring product launch recently marks the beginning of a new era for Smart Rings. Smart Rings have entered the mainstream, and market growth will accelerate accordingly. ___ has published a white paper to analyze the market and help companies define their go-to-market strategy!"

What makes this even more intriguing is if you replace "Smart Rings" with any other high-tech term, this release makes just as much sense. Well, as much sense as this press release actually makes—which is none. They missed the fact that a general tech audience has no clue what they are talking about. They defined nothing. Explained nothing. Never answered why this announcement was important to their audience.

And wasted a lot of people's time.

It is society that "makes" color, defines it, gives it its meaning, constructs its codes and values, establishes its uses, and determines whether it is acceptable or not. The artist, the intellectual, human biology, and even nature are ultimately irrelevant to this process of ascribing meaning to color. The issues surrounding color are above all social issues because human beings live in society and not in solitude.[1]

MICHEL PASTOUREAU
FRENCH PROFESSOR OF MEDIEVAL HISTORY

CHAPTER 5
PERSUASIVE COLORS

CHAPTER GOALS

The goals of this chapter are to:

- Explain why understanding color is important
- Present a brief history of color and the impact of Sir Isaac Newton
- Discuss the "meaning" of color
- Present a simple color model and define key terms
- Illustrate a variety of color techniques
- Suggest ways to pick the "right" color for your project

[1] (Pastoureau, 2001, p. 10)

THE ROLE OF COLOR is vast, visceral, and vital to our lives and emotions. We "color our opinions," "emotions color our thoughts," or "we add color to a story." We hassle with red tape, celebrate our green thumb, or brag about a white knight appearing out of the blue.

Color attracts attention.

It is impossible to have a conversation these days without color featuring prominently. Medieval historian Michel Pastoureau writes, "In all cultures, color's primary function is to classify, mark, announce, connect, or divide."[2] As an example, here in the United States, the terms "red" and "blue" represent far more than a simple name for a color. Like society, our perception of color evolves over time. In this chapter, I want to look at color from the perspective of persuasion.

Color often provides meaning and emotion, but, above all, color attracts attention.

Nothing creates a faster emotional impact than color (**FIGURE 5.1**). The use of color is essential to almost all visual messages. But, color is complex, and as you'll discover, different societies interpret the same color differently.

FIGURE 5.1 We recognize these based on shape and color, long before we read the text. In fact, the merge sign doesn't even need text to convey its meaning! (Image Credit: Wikimedia Foundation, Inc.)

Martin Lindstrom is a researcher who spent three years and seven million dollars studying neuromarketing. In his book *Buyology*, he wrote, "A study carried out by the Seoul International Color Expo found that color goes so far as to increase brand recognition by up to 80 percent. When asked to approximate the importance of color when buying products, 84.7 percent of total respondents claimed that color amounted to more than half the criterion they consider when they're choosing a brand. Other studies have shown that when people make a subconscious judgment about a person, environment, or product within ninety seconds, between 62 and 90 percent of the assessment is based on color alone.[3]"

This chapter provides a greater understanding of color and how to use it to improve our presentations, images, and videos.

A Brief History of Color

To set the scene, let's start with a brief history of color. Here, I'm indebted to *Blue*, a fascinating book written by historian Michel Pastoureau, on the history and use of the color blue from prehistory to today. Just like the film *Helvetica* that I mentioned in Chapter 4, "Persuasive Fonts," there's more to even a single color than you might think.

[2] (Pastoureau, 2001, p. 10)
[3] (Lindstrom, 2008, p. 155)

Prehistoric Colors

Back in the dawn of human history, in the Paleolithic and Neolithic periods, artists lived in a world of reds, oranges, and black. It wasn't that blues or greens or yellows weren't abundant in nature. They were. But they didn't exist as pigments or paint.

Back then, the three fundamental colors were red, white, and black. Other colors existed but were not primary (pun intended) in people's lives. For example, neither Greek nor Roman languages had a word that specifically described the color blue. For the Romans, blue was a "barbarian's color." It was a color to be distrusted and avoided.[4] Why they felt that way is unknown today.

Dyeing cloth began sometime between the sixth and fourth centuries BC. (All the ancient cloth fragments we've found so far were dyed red, from the lightest pinks to the darkest purples.) Ancient civilizations described a color in terms of its lightness and its amount, what we would call "grayscale" and "saturation" today.

> ### How We See Color
>
> According to the latest research, the colors we see are reflected off an object. An apple, for example, absorbs all the colors *except* red, which it reflects to our eye. So, while we see the apple as "red," it could be argued that it is every color *except* red. This weird fact does not stop me from happily eating apples, however.

This emphasis on red, white, and black extended into the high Middle Ages (1000–1300 AD). Ignored by nobles, blue cloth was colored from either woad (a cabbage-like plant) or indigo (a plant in the pea family) and worn only by peasants. In the early 1200s Pope Innocent III, the most significant pope of the Middle Ages, described the liturgical value of the three primary colors: "White symbolizes purity...red evokes the blood spilled by Christ...black [is] the color of grief and penance."[5] The blue robe indicating the Virgin Mary had not yet made an appearance.

Here's another interesting color fact: White was considered to have *two* opposite colors, black and red. Until the Renaissance, black was somber, and red was dense, while white was the opposite of each.[6] (Dense meant the "amount" of color, what we now call "saturation." Red has lots of color, while white appears to have none.)

Then, Along Comes Isaac Newton

In 1666, Isaac Newton shattered millennia of color theory by punching a hole in his window shade and cast a beam of sunlight onto a prism. The light spread into what he described as a "spectrum" of seven different colors. The number seven was significant, Newton felt, because since there were seven

[4] (Pastoureau, 2001, p. 26)
[5] (Pastoureau, 2001, p. 40)
[6] (Pastoureau, 2001, p. 18)

whole notes in music (A–G), there must also be seven pure colors, which he described as red, orange, yellow, green, blue, indigo, and violet.

While splitting light through a prism was not new—in fact, the Church at the time taught that splitting light was the same as breaking it—Newton's research went one step further. He took a second prism, captured the spectrum, and showed that it recombined back into white light. This recombination of light was his first big insight: light was formed of component parts, which could be easily separated and recombined. This discovery was both revolutionary and heretical!

Newton's prism experiments showed that light, rather than being a pure, single color, was actually composed of a wealth of separate colors. While indigo was dropped when it was realized that there was no relationship between musical notes and colors, the rest of Newton's observations revolutionized the way we think of color.

What really galvanized color thinking was Newton's second insight: placing the colors in a circle, or a "wheel" (**FIGURE 5.2**). Although the spectrum of light is linear, with low frequency (red) on one side and high frequency (blue) on the other, Newton's tied these two ends together to form a circle. In fact, the "color wheel" he designed 400 years ago remains very close to the way we see and select colors today (**FIGURE 5.3**).

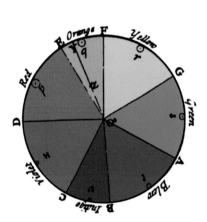

FIGURE 5.2 Sir Isaac Newton's sketch of the seven basic colors of white light, which he placed into a color wheel. (I added the colors.)

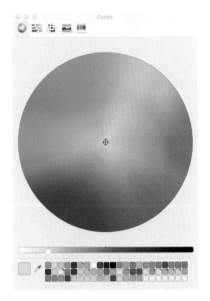

FIGURE 5.3 A modern "Newton color wheel"—the macOS Colors window.

Frequencies of Light

Just as sound is composed of vibrating waves of air, light is composed of varying waves of light. These "frequencies" determine the color of light. Measured in trillions of waves per second, lower frequencies shift a color toward red, while higher frequencies shift a color toward blue. All other colors exist between these two extremes.

Color Also Has Temperature

We often speak of a color as "warm" or "cool." What we are referring to is its color "temperature." Color temperature represents the color of a black reference object as it is heated.

While our eyes instantly adapt to different colors of light and are good at figuring out what's white, digital cameras are not as good as our eyes. Often, the color of an image will vary depending upon whether it is lit by the sun at noon (bluish), interior house lights (warmish), or sunset (really warmish).

In practice, color temperature is meaningful only for light sources that range from red to orange to yellow to white to bluish white (**FIGURE 5.4**). Color temperature is conventionally expressed in kelvins, using the symbol "K," which indicates how much you would need to heat a "reference black body" to get it to glow at a specific color.

While Kelvin is a measure of temperature in degrees above absolute zero, unlike the degree Fahrenheit and degree Celsius, the Kelvin is not referred to or written as a degree. So when we refer to the color of light, we simply use "K," with a space between the number and the letter.

Here are some fun facts about color:

- What we would consider "white" light is around 6500 K.
- The effective color temperature of the sun is about 5780 K.
- Most natural warm-colored light sources emit significant infrared radiation, which may be why we call them "warm."
- Sunrise/sunset is around 1850 K.
- Interior household lights are around 2700 K.
- The sun at noon is around 5500 K.

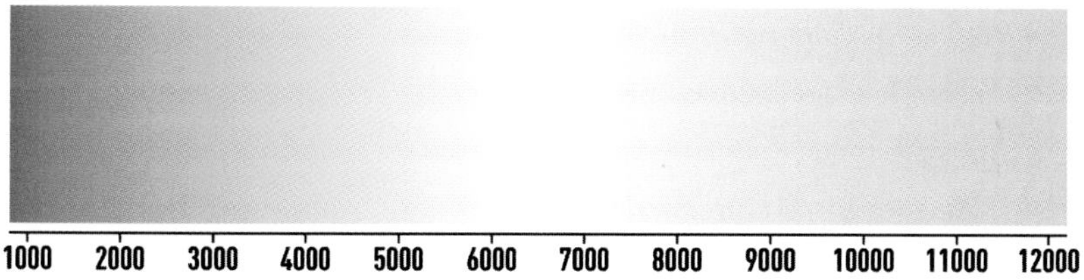
1000
2000
3000
4000
5000
6000
7000
8000
9000
10000
11000
12000

FIGURE 5.4 This represents the range of color temperatures of white light, as measured in degrees Kelvin. (Image courtesy of Bhutajata / CC BY-SA)

- Overcast skies are around 6500 K.
- Clear blue sky is around 15,000 K.
- Color temperatures over 5000 K are called "cool colors" (bluish), while lower color temperatures (2700–3000 K) are called "warm colors" (yellowish).
- Bizarre fact: The temperature of a "warm" light is cooler than the temperature of a "cool" light.
- It does not make sense to speak of the color *temperature* of a green or purple light. You can't heat something to make it radiate green, though both these colors are still part of the light spectrum.

In both photography and video, you'll hear terms like "tungsten" (3200 K) and "daylight" (5500 K). Both are used to describe white light in terms of color temperature.

If you take pictures that have an orange or blue color cast, you are dealing with a camera that was not set to the proper color temperature before taking the picture.

The Meaning of Color

While looking back at the history of color is interesting, it begs the much larger question: How can we use color to guide the emotions of our viewers? In other words, what does color mean? This is an easy question to ask yet impossible to answer in the absolute.

Michel Pastoureau, again, explains why: "Color is first and foremost a social phenomenon. There is no transcultural truth to color perception, despite what many books based on poorly grasped neurobiology or—even worse—on pseudo-esoteric pop psychology would have us believe."[7]

When I first outlined this chapter, my goals were to provide a simple guide to choosing the "right color for your project." How hard could that be? For example, the SAE Institute shares this color guidance with their artists:

- Red is the color of fire and blood.
- Yellow is associated with the sun, the color of optimism and joy.
- Orange stands for energy and happiness.
- Blue is cool and calming and stands for intelligence and creativity.
- Green is the color of nature and the environment.

[7] (Pastoureau, 2001, p. 7)

- Purple is for royalty and ceremony.
- White represents power and elegance and death.
- Black represents power and elegance and death.[8]

I was doing OK until I read the descriptions for white and black. How can two different colors represent the same emotions? (For that matter, the descriptions of yellow and orange are pretty darn close as well.) This brought me back to Michel's quote at the start of this chapter: the meaning of color is social, not absolute.

Cognitive scientist Don Norman and his colleagues investigated whether there were genetically programmed responses to specific situations, colors, or objects. What they found was that while emotional responses to different colors vary by culture, there are "situations that, through evolutionary history...give rise to positive affect." **TABLE 5.1** shows a partial list of the automatic positive or negative responses that they found.

This is not to say that colors don't have meaning, simply that meanings are not universal; they are societal.

TABLE 5.1 Universal Reactions to Colors	
POSITIVE	**NEGATIVE**
Bright, highly saturated hues	Sudden, unexpected loud sounds or bright lights
Harmonious music and sounds	Empty, flat terrain (deserts)
Smiling faces	Harsh, abrupt sounds
Rhythmic beats	Grating and discordant sounds
"Attractive" people	Crowds of people[9]
Symmetrical objects	
Rounded, smooth objects	

No specific colors made this list, because each society determines how to interpret color, but the *shapes* that display colors did make the list. Like the stop sign at the beginning of this chapter, shapes and colors are often combined. It is also worth noting that many of the pleasing responses Dr. Norman reported are based on what we see, while negative results are often based on what we hear. We can use these guidelines in creating messages that are more likely to trigger the appropriate response in our audience.

Graphic designers Kevin Budelmann, Yang Kim, and Curt Wozniak created a series of design principles for basic branding in their book, *Brand Identity Essentials*. They wrote, "Clinical and anecdotal tests on color psychology and emotion have led to the development of widely accepted theories about

[8] https://alumni.sae.edu/2016/03/08/what-do-colors-mean-and-represent/
[9] (I omitted from his list things that we touch or smell. Norman, 2004, p. 31)

We can't talk about the "meaning" of a color unless we also talk about the social environment in which that color is being viewed.

color. …But the power of certain colors changes over time and across cultures. One cannot deny the influence of fashion industry trends on color choices. …Culture also plays a role in how colors are interpreted. The obvious example: In Western cultures, people wear black to funerals, while in Eastern cultures mourners wear white. The culture connotations of color are often learned and permeate a market."[10]

In India, red may indicate a woman is getting married, while in South Africa, it means mourning. In Thailand, red is the color for Sunday and Surya, a Thai god, while in China, red symbolizes the New Year and good luck.

Or consider blue. In the West, blue is associated with a feeling of melancholy yet is also calming and soothing. Blue is masculine in the West, but feminine in China. In the Middle East, blue means safety and protection, while in Latin American countries, blue represents hope, good health, and the Virgin Mary. Blue also has religious overtones in Judaism and Hinduism.[11]

We can't talk about the "meaning" of a color unless we also talk about the social environment in which that color is being viewed. This is not to say that colors don't have meanings, simply that meanings are not universal. And if our goal is persuasion, we need to take the social value of color into account in creating our messages.

THE Dress

On Feb. 26, 2015, this dress exploded across the Internet. Was it white and gold or blue and black? Tens of millions of people expressed their opinion.

"The dress," as it's now called, is a photograph of a dress that caused more than 10 million tweets. Millions of viewers debated and from that emerged a bigger discussion of how we perceive color. The answer is that the original dress, made by Roman Originals, is black and blue, and the photo is somewhat overexposed.[12] (Ah, no, I didn't see those colors initially, either.)

So, which is it: white and gold—or blue and black?

[10] (Budelmann, Kim, & Wozniak, 2010, p. 16)
[11] https://www.huffpost.com/entry/what-colors-mean-in-other_b_9078674}
[12] https://en.wikipedia.org/wiki/The_dress, (Image Credit: Wikimedia Foundation, Inc. / Cecilia Bleasdale)

A Simple Color Model

I haven't lost sight of our ultimate goal: to learn how to use colors to persuade others. But, since color is firmly rooted in history, culture, art, *and* technology, we need to find a more organized way to talk about it technically.

It is hard to describe the full range of color that we see. **FIGURE 5.5** is an analogy that often helps my students. Imagine that all the colors we can see are stored inside a grapefruit. To help us find them, let's get them organized.

FIGURE 5.5 The easiest way to think about all the colors we can see is to imagine them stored in a grapefruit, sorted along three axes. (Image Credit: Wikimedia Foundation, Inc./ Evan-Amos)

Sort the colors so the darker ones move toward the bottom (south pole) of the grapefruit, while the lighter ones move toward the north pole. Next, sort the colors by hue so those that are reddish clump together, as do blue and green, and so on. Finally, imagine them sorted so those that contain the most color are located near the outside edge, while those that have the least color are located near the center.

We've just created a color space. (I'll get back to these three axes in a few pages.)

Let's cut the grapefruit at the equator (**FIGURE 5.6**) and look at the cross section. We see all the colors of Newton's rainbow displayed within it. As we move around the circle ("change the angle"), the hue changes. As we move out from the center, the amount of color ("saturation") increases from gray to fully saturated. And brightness shades smoothly from pure black at the south pole to pure white at the north pole, creating a gradient.

FIGURE 5.6 Cut the grapefruit at the equator and look at the cross section. Color hue varies by angle, while color amount (saturation) varies by its distance from the center.

While the actual color space our eyes see is not perfectly round, this grapefruit serves as a useful analogy to describe and manage color for both still and moving images.

Digital Color Terms

With our color space in mind, let's define some key color terms that are relevant in our computer age.

Pixel. The smallest component of any digital image. (It's short for "picture element.") In today's digital images and video, pixels are always square and always contain one, and only one, color.

Bitmap. A rectangular grid that defines how pixels are arranged. All digital images are displayed as bitmaps. There is one bitmap per image.

Resolution. When applied to digital images, the number of pixels in an image. All digital images are rectangular bitmaps. When describing a bitmap, we refer to the horizontal dimension first. For example, 200 x 400 pixels describes the size of a web ad, which is 200 pixels wide by 400 pixels high for a total of 80,000 pixels. 1920 x 1080 pixels describes the size of a high-definition video frame; it means 1,920 pixels wide by 1,080 pixels high, for a total of 2,073,600 pixels. Compared to images destined for print, which are often measured in tens of millions of pixels, digital images for use on the web or in video have limited resolution but a virtually limitless range of colors and translucency.

DPI/PPI/resolution. Dots per inch/pixels per inch. These terms define *pixel density*, how many pixels per inch a printer uses to reproduce an image. These settings apply *only* when an image is printed. When creating images for web or video, these settings are ignored due to the wide variation in computer monitors. By default, we set this to 72.

Grayscale. The brightness of a color on a range between black (dark) and white (light). Pink has a light grayscale value. Maroon has a dark grayscale value. The easiest way to think of grayscale values is to think of images in a black-and-white photo. All you see are shades of gray, not colors.

Contrast. The grayscale difference between two pixels or images.

Color. This is a catchall term for all the different characteristics that make an object seem red or green.

Chroma. The hue and saturation values of a color. "Chroma" is shorthand, so we don't need to keep saying "hue" and "saturation."

Locating Colors Precisely

Remember the grapefruit "color space" analogy in Figure 5.5? Using it, we can now precisely locate any color using only three numbers:

- **Grayscale.** The vertical axis represents the grayscale, the lightness value of a color.
- **Saturation.** The distance from the center represents saturation, the amount of a color.
- **Hue.** The angle around the edge of the sphere represents hue, the shade of a color.

We will see this three axes concept again in a few pages when we discuss measuring color and Vectorscopes.

Hue. A component of chroma, representing the shade of color; purple, for example.

Saturation. A component of chroma, representing the amount of a color. Gray has no saturation; neon yellow has a lot of saturation.

Saturation is often a hard concept to understand. In these cyan color chips (**FIGURE 5.7**), both sides have *exactly* the same hue (202°), but not the same saturation. The chip on the left is saturated 15 percent, while the chip on the right is saturated 70 percent. Both are cyan, but there is more color, more saturation, in the chip on the right.

FIGURE 5.7 The color on the left is the same hue as the color on the right (202°). The difference is saturation, the amount of color in each square.

RGB. The universal color space of digital images that describes a color in terms of how much red, green, and blue it contains. We'll work exclusively with RGB images for all digital media and the examples in this book.

CMYK. The universal color space of printed images that describes a color in terms of how much cyan (C), magenta (M), yellow (Y), and black (K) it contains. Essentially, CMYK is the inverse of RGB.

Primary colors. For RGB images, these are red, green, and blue (**FIGURE 5.8**).

Secondary colors. For RGB images, these are the exact opposite of each primary color: cyan, magenta, and yellow.

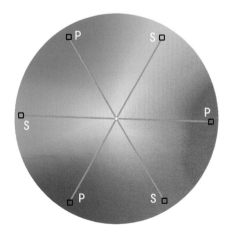

FIGURE 5.8 Primary (P) colors for RGB images are red, green, and blue. Secondary (S) colors for RGB are their direct opposite: cyan, magenta, and yellow.

FIGURE 5.9 When combined, RGB primary colors (top) *add* to create white, while CMYK colors *subtract* to create black.

In RGB color space, adding opposite colors (Figure 5.8) equals gray. We'll use this characteristic to remove color casts in stills and video later in this book.

Next are three more technical terms.

Additive. A characteristic of RGB colors that when they are combined in equal amounts, add to create light gray (or white) (**FIGURE 5.9**).

Subtractive. A characteristic of CMYK colors that when they are combined in equal amounts, they subtract to create dark gray or black. (Because this color subtraction doesn't work perfectly, black is often added during printing to convert a dark gray to black.)

Bit-depth. This number, illustrated in **FIGURE 5.10**, describes the number of shades of color or gray a pixel can have. Expressed as a power of 2, 8-bit color supports up to 256 shades (2^8) of gray (or a color). 10-bit color (2^{10}) supports up to 1,024 shades. The lower the bit depth, the more "steppy" gradients will look. All web images are 8-bit.

FIGURE 5.10 This is an exaggerated illustration of the difference between bit depth values of 8-bit (top) and 10-bit. As bit depth increases, the smoothness of a gradient improves.

Grayscale

Grayscale is a hard concept for many of us to grasp since we spend our whole lives looking at the world in color. Grayscale simply means that portion of an image that isn't color. Grayscale displays an image as shades of gray. Most of us would call **FIGURE 5.11** a black-and-white photo.

FIGURE 5.11 A color image (left) versus grayscale (though you may also call this a "black-and-white" image.)

FIGURE 5.12 Compare the emotional differences: The color image feels "real," the grayscale feels desolate, while the sepia feels old, exhausted, and hopeless. (Image Credit: Gerd Altmann / pexels.com)

As well, in Figure 5.11, it is easy to see which pixels are lighter than others. As you look at the color image, you can see how colors with darker grayscale values have darker colors as well.

There is also an emotional context to black-and-white versus color. Compare the differences in **FIGURE 5.12**. Removing the color increased the feeling of loneliness and desolation. This is one reason why, when comparing two products or people, the one to be viewed in a negative light is often in black-and-white. The sepia version on the far right feels older and exhausted, as though it's an image from the Great Depression.

To get the best results when creating a sepia version of a color image, first convert the image to black-and-white then add the sepia tint.

Yet, in all three cases, the image is the same—all that changed was the color treatment. As I look at these three images again, the black-and-white seems the most believable in conveying sadness. The color version feels too cheerful to be this depressed, while the sepia version takes me back to a different time.

Color vs. Contrast

Understanding grayscale can also help us with readability, our first goal in creating any image. There are two concepts that can help: color and contrast. I'm using "color" to mean the amount of color in an object, while contrast is the amount of variation between light and dark.

In **FIGURE 5.13**, the color of the four words is the *exactly* the same! But they are different shades of gray; the top word is darker than the bottom. (This reminds me of Molly Bang's observations in Chapter 2, "Persuasive Visuals": Color has "weight." The darker color feels "heavier," as though it should be on the bottom, while the lighter color should float to the top.)

FIGURE 5.13 This illustrates contrast. All four of these words are *exactly* the same color (meaning identical hue and saturation) but not the same grayscale.

Also, notice that the same color word looks different against a white background versus a black one. This becomes even more true when you put text of one color on top of a background with a different color. Sometimes, they look great; other times they, ah, don't.

In **FIGURE 5.14**, the grayscale and saturation values are exactly the same; only the hue is different. Notice how the colors seem to change as the background changes. Also, on the right, I removed the color and the text became almost impossible to read. Actually, I cheated. There's a 2 percent difference in grayscale between the text and the background. If they exactly matched, you wouldn't see the words at all.

The point I'm making is that simply changing a color may not improve readability. We may need to change color *and* grayscale, as well as pay attention to the background behind the text.

Consider Color Blindness

Color blindness affects about 1 in 12 men, and 1 in 200 women, according to Color Blind Awareness. This means when you are picking colors, you want to choose colors that vary both in color and in grayscale. Without variation in grayscale between text and background, if a member of your audience is color blind, it may be impossible for them to read your text. It's analogous to trying to read the text on the right side of Figure 5.14—it is very difficult.

FIGURE 5.14 The grayscale and saturation values are *exactly* the same, but not the hue. The right image illustrates the grayscale value with no color.

In general, white text is preferred for titles (harking back to the Six Priorities), which means that we need to be careful to put it on a contrasting background or use a drop shadow. As well, you want to vary both color and contrast when working with different colors in the same image.

Let's examine how the background affects readability. In Figures 5.13 and 5.14, we saw how the perception of color changes based on the background. **FIGURE 5.15** is another example. Lighter text washes out against white but "punches" against black. The contrasting color (setting the background to the opposite color of the text on the color wheel) makes text easily readable. The variegated background, with similar grayscale values to the text, makes the words virtually unreadable.

But, adding a drop shadow (**FIGURE 5.16**) makes all the difference. Unreadable text suddenly becomes readable. Since we rarely have full control over the image background, drop shadows are a good habit to get into. Drop shadows don't need to be obvious. We want the viewer to say, "Nice text!" not "Cool drop shadow!" As always, readability is key.

One other note. As we discussed in Chapter 4, "Persuasive Fonts," drop shadows should be applied to all title text. Especially for video, they will also improve the readability of body text.

Drop shadows don't need to be big and obvious. We want the viewer to say, "Nice text!" and not "Cool drop shadow!"

FIGURE 5.15 Look at how the color and readability change as the background changes.

FIGURE 5.16 Adding a drop shadow to the text on left and right solves the readability problem.

What Color Is an Iguana?

If I asked you, "Quick! What color is an iguana?" What would you say? Pink, green, gray, blue...? If I showed just about anyone a photo of a red iguana and said, "This is a photo of the extremely rare crimson iguana!" They'd say, "Cool!" Because, after all, what color is an iguana? We don't have the experience to know *for sure* the colors of every possible variation of iguana that would prove me right or wrong. (Yes, there are iguana experts, but there are not a lot of them; and actually, there *is* a Galapagos pink land iguana. Cool! Who knew?)

The point I'm making is that for many images, the exact color of something is not particularly relevant because no one knows the "true" color anyway. The color of an object is whatever we tell the viewer it is.

But this acceptance is *not* true for three specific and important colors that are deeply ingrained in our memory: green grass, blue sky, and skin tone. These three are called "memory colors." We may not be able to describe them in words, but we can tell instantly if they are off, even a little bit.

Show me an image of an iguana and I'll cut you some slack if you say it is pink. Show me an image of someone's pink lawn and I'll know instantly that I'm looking at fake grass.

For example, in **FIGURE 5.17** we instantly recognize the accuracy of the color of the blue sky, even though it isn't really blue. It's much closer to cyan. If I showed you a photo with a blue sky similar to the blue color chip, you'd instantly know the sky was fake.

When there's a reasonable amount of doubt, the viewer will tend to believe you. But not if you screw up one of the memory colors.

FIGURE 5.17 "Blue skies" aren't blue. The square is blue. The sky is cyan.

The Special Case of Skin Color

This discussion of memory colors brings us to the special case of skin color—also called "skin tone." Skin is a complex surface in terms of color, luminosity, and reflection. What gives skin its special color is the red blood coursing just underneath the surface. In fact, it's more accurate to say that all of us have the same *color*, but we don't have the same *grayscale*.

Alexis Van Hurkman in his outstanding book, *Color Correction Handbook*, analyzed variations of skin tone, including "self-identified Asian, African-American, Indian (Indian subcontinent), and Caucasian individuals, with additional sampling for pale/red-haired subjects and tanned members of each population. All the skin samples in **FIGURE 5.18** were previously color-corrected so they present a nonscientific survey of various photographers' ideals for appealing complexion."[13] You can see the results in Figure 5.18.

[13] (Hurkman, 2011, p. 308)

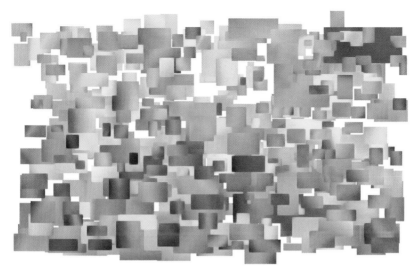

FIGURE 5.18 Across cultures, skin-tone values are limited to a very small range of "color," but a wide range of grayscale. (Image Credit: Hurkman, A. V. (2011, p. 308). *Color Correction Handbook.* Peachpit Press.)

Think about getting cleaned up in the morning when a piece of skin falls off. You first check to make sure it wasn't something vital, then you notice that it has a gray color. Skin is essentially translucent. It's only when you combine it with the color of blood under skin does it achieve its familiar color.

Again, skin tone is a memory color. We all know what "normal" skin looks like. This is not to say there isn't variation, but if you are trying to make someone look "normal," you have a very narrow color range to fit within.

Skin tone is one of the memory colors that we need to get right for our audience to believe our message. "If the color of their skin isn't right," the viewer thinks, "how can I possibly believe anything else in this image?" (I'll have more about skin color later in this chapter.)

How We Measure Color

This section gets a bit technical, but I want to keep our discussion of color in one place to make it easy to refer to later in the book. While there are lots of instruments that can measure color precisely, there are four we use the most.

- The macOS Colors window
- Histogram (principally used for stills)
- Waveform Monitor (principally used for video)
- Vectorscope (principally used for video, too)

FIGURE 5.19 Three views of the macOS Colors window: Color wheel, Spectrum, and RGB values.

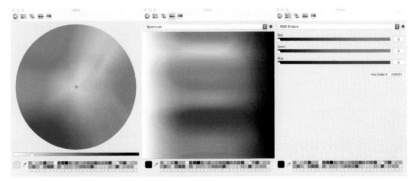

The first place many of us go to choose a color is the macOS Colors window, shown in **FIGURE 5.19**. This enables choosing colors by eye or entering numerical values to match colors precisely. While different applications may have variations on this interface, the operation is the same: Pick the grayscale and color values you need by either dragging a slider or entering specific values. The macOS Colors window also provides spaces to store frequently used colors at the bottom.

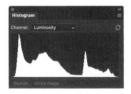

FIGURE 5.20 The Histogram shows the range of grayscale values in an image. Black values on the left shading to white on the right. (Screenshot from Adobe Photoshop.)

The Histogram, **FIGURE 5.20**, is used to measure and adjust grayscale values for still images; it can be used for video, but video tends to use other tools. The display places black on the left, shading to white on the right. Between them are all the different shades of gray. The height of the white curves in the graph represents the number of pixels at a specific grayscale value. (Higher peaks mean more pixels share that grayscale value.)

FIGURE 5.21 shows a Histogram used for video. (Both Histograms display results for the same image—the mountain photo shown in the Sidebar.) The grayscale values for the image are the same, but the displays are not because Photoshop scales the display for still images, while Final Cut (and Premiere) scales the display for video.

The Waveform Monitor will be new to many. As we move to video, the Histogram becomes less useful. Instead, we use the Waveform Monitor to describe the grayscale values in an image. Unlike the Histogram, left to right in the Waveform Monitor represents left to right in the image, while vertical values represent the brightness of each pixel (higher is brighter). The brighter the "trace"—that is, the glowing gray portion of the image—the more pixels that share a specific grayscale value. In **FIGURE 5.22**, the trace is brighter near the bottom, which indicates that there are more dark pixels than lighter pixels in this image.

The Vectorscope (**FIGURE 5.23**) displays the same image as the other scopes, but in a different manner: The Vectorscope displays color values as the

Waveform Monitor displays grayscale values. Different angles in the Vector-scope represent different hues. Saturation increases as we move out from the center. It's identical in concept to the grapefruit color wheel, though the color angles have been rotated. (Red is located in the upper left, blue is in the right center, and green is in the lower left.)

FIGURE 5.21 This is the Histogram display from Apple Final Cut Pro X. It measures the same image but scales the display to meet video standards.

FIGURE 5.22 This is the Waveform Monitor. This shows grayscale values but not color. (Screenshot from Apple Final Cut Pro X.)

FIGURE 5.23 The same image, displayed on a Vectorscope. This shows color values but not grayscale. (Screenshot from Apple Final Cut Pro X.)

The Histogram is primarily used for still images, while both the Waveform Monitor and the Vectorscope are used more in video. The Vectorscope will be extremely helpful in finding and fixing color problems.

While the tools we use to adjust color vary between still and video images, the concept is the same: Adjust the grayscale values first then adjust color values. I'll cover how we make these adjustments in future chapters.

Measuring Skin Color

We use the Vectorscope to measure and verify color settings. A common use is to measure skin color, as shown. Let's take a look at the skin swatches we first saw earlier. If you look carefully at the Vectorscope image, on the left, you'll see all the skin colors fall into a *very* narrow range of hues (angle), though a fairly large range of saturation (distance from the center). As I tell my students, "We are all the same color—but not the same grayscale."

Look even more closely and you'll see a thin line radiating up to the left in the Vectorscope. This is called the "skin-tone line" and represents the hue of normal human skin under "normal" lighting. (By normal, I mean normal daylight or studio interview lighting, not some dimly lit disco with flashing neon red lights.)

If skin tones fall very close to this line, as they do in this example, you know skin color is accurate. If they wander, you know that you have a color cast problem that needs to be fixed.

Note how a wide variety of skin types fits within a very small area of the Vectorscope.

KEY POINTS

Color is a complex subject. Here's what I want you to remember:

- Color attracts attention.
- Color is a powerful emotional driver, but the interpretation varies from one culture to another.
- Because our goal is persuasion, pick colors that enhance your message.
- The Six Priorities remind us that "different" attracts the eye.
- Pick colors that make your elements readable.
- Pick colors that have different contrast values.
- Some viewers are color blind; don't base your entire message on a single color or colors with the same grayscale value.
- Don't get fancy—fewer colors and styles are always better.
- Ask others their opinions to make sure your perceptions of how colors are interpreted matches with your audience.

PRACTICE PERSUASION

Extend the exercise we did in the previous chapter with fonts to colors. Using a word processor, set this headline in different colors and against different color backgrounds. How does the choice of color and contrast affect the perception, emotion, and urgency of the message?

New Look! New Colors! New Style!

PERSUASION P-O-V

SEEING WITHOUT KNOWING

Jean Detheux
Painter/filmmaker

As a painter and filmmaker, I am often surprised by how little people who deal with visual communications (including video game design) seem to know about first-person perception.

Indeed, the way most people operate is not rooted in how we experience "the real." It is most often stuck in a model of the world that posits photo-realism as a valid representation of how we experience visual "reality."

(Image Credit: Jean Detheux, Painter filmmaker)

That misconception has a significant impact on everything we do; it conditions the scope of the field in which we will contemplate doing whatever we will undertake. Yet, if one explores first-person perception, one quickly discovers a world that does not conform at all to the photorealism model.

"The eye works like a camera and we all see the same things" is a deception, a lie, one of the most prominent aspects of what has been called "the fallacy of misplaced concreteness."

Edmund Husserl, the father of phenomenology, called this "ontologizing." Ontologizing posits "the world" as objective, finite, constant, and our experience thereof as merely subjective, transitory.

But unlike the most common worldview would have us believe, we do not start from an "objective" view of "the world"; our primary connection with it is subjective, through and through.

Freely paraphrasing Maurice Merleau-Ponty in his masterful *Phenomenology of Perception*, our knowledge of the world, including scientific knowledge, relies on our particular point of view. ("Each one of us is a different point of view in the world.")

Perception works in mysterious ways, but if one starts to pay attention to how one's perception works, it starts to open up like a book that one knew well but had (almost) forgotten about.

> *To see we must forget the name of the thing we are looking at.*
> —CLAUDE MONET

After all, we are not talking here about a perception that is remote from one's very being, a perception one would have to learn then apply to one's experience; we are indeed talking about what constitutes and has always constituted our first, most immediate, and intimate rapport with "the world."

It is the perception we often overlook, that which we utterly take for granted and want to do something "with." Yet, if we could simply "see" it, on its terms, and how it comes about, we likely would no longer want to "do something with" it; it is such a fertile ground, some of us will not have enough of a lifetime to explore all that it offers.

Many painters, like Claude Monet, know well that they must "forget" the name of a thing to truly see it. They know that investigating preverbal perception is as sophisticated and worthy an area of exploration as it is utterly basic; in the most noble sense of the word.

While I could show many examples of works by painters who understood this well, I will simply compare the image shown at the start of this article, with the image in full: a painting by Rembrandt.

Look at the difference in this painting between the light and detail of the face, compared to the barely sensed brushes and palette in his hand. Artists call this "sense-giving" (the face) and "sense-receiving" (the hand and palette). This is a beautiful example of central and peripheral vision, of the origin of abstraction in perception, the abstraction that links the greatest artists from cave painting to today's art. The face is the central focus, while the rest of the image, senses peripherally, gives the face and total painting context.

When building a visual presentation, it can be easy to overlook the fact that, in spite of the content of that presentation, its form also plays an immense role. That form operates on at least two axes: the form of the individual "slides," seen as still images or frames, and the form of the presentation unfolding in

"The eye works like a camera and we all see the same things" is a deception, a lie.

—JEAN DETHEUX

time. In other words, both the information on each slide and the context of how they are presented over time are important.

Much of what one can learn about the behavior of elements on one axis applies to both, something that surprised (and delighted) me when I was forced many years ago to leave the world of painting (sudden and severe allergies to natural media) for that of the digital workflow, which quickly opened up the time dimension for me.

When building a visual presentation, it can be easy to overlook the fact that, in spite of the content of that presentation, its form also plays an immense role.

Elements at play are numerous, one of the most important being the qualitative relationship between "sense-giving" (Rembrandt's face/gaze in the full painting) and "sense-receiving" elements (Rembrandt's hand and palette shown at the beginning of this story).

Not all elements are equal and not all carry the same amount of meaning, but the least important ones are needed by the most important if they are to deliver a clear and lasting "message," a content they build and have to build together, hence their qualitative relationship.

And those different elements are not always present, visible, at the same time: In something like a Keynote or PowerPoint presentation, they may be present on different slides and yet communicate across time through the meaning they are building together (and this, of course, occurs in the viewer's "mind").

An element barely mentioned at one point in time may gain significant meaning retroactively when another element makes reference to it later, giving it increased meaning, meaning it did not seem to have when first presented (possibly creating in the viewers something akin to an "aha!" experience).

In the field of painting, this dialogue in space between elements has been called "echoing shapes," and I was very pleasantly surprised when I discovered, after I made the switch from painting to movies, that these echoing shapes also begged to be "used" in film as well and are very powerful.

In keeping with the form of a visual presentation, there are many other elements that are often ignored by the people making the presentations, and I was struck years ago by one in particular: the material presented on one slide is often placed in such a way that when the switch to the next slide occurs, there are "jumps" in the content thus revealed and often, this jump supersedes in the viewers' attention the content the presentation hoped to convey.

The placement of elements that carry from moment to moment (from slide/frame/scene to slide/frame/scene) has not been shaped to participate fully in the unfolding of the content.

These techniques are not rocket science. Many years ago, I taught them to my 8-year-old son who was building a Keynote presentation for his school, and he really got it. It was amusing to see my young son competently using elements many of my professional colleagues often overlooked.

Viewing and viewing and viewing again previews of one's presentation should start revealing weaknesses (and strengths), and any new knowledge gained this way will build up an awareness of the attitude needed to access this mode of perception with some confidence.

"Seeing without knowing" is not negating knowledge; it is a way of being that provides access to the very source of all knowledge.

Perception works in mysterious ways, but if one starts to pay attention to how one's perception works, it opens like a book one has forgotten about.

—JEAN DETHEUX

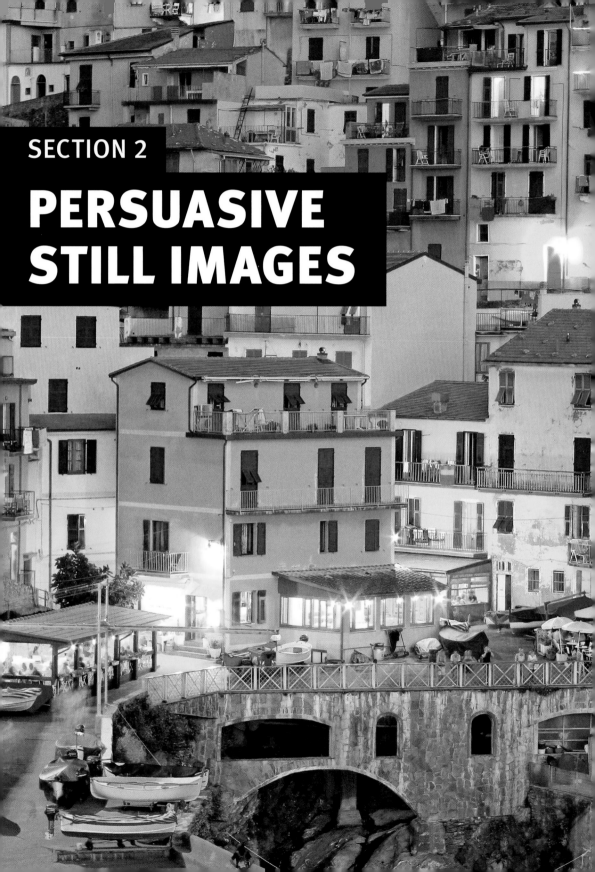

SECTION 2

PERSUASIVE STILL IMAGES

SECTION GOALS

This book is about persuasion. More importantly, it's about specific techniques we can use to persuade others. At the heart of everything we do is the goal of attracting and capturing the eye of the viewer. The Six Priorities remain a critically important tool to determining which element in our design will attract the eye first, then second, then third.

So far, we've covered a lot of theory; now we put that theory into practice by looking at still images. Starting with the simplest of stills, a business presentation, we expand from that into photography, image editing, and, finally, creating images that can't be recorded by a camera.

The chapters in this section are:

- **Chapter 6: Persuasive Presentations.** How to make a persuasive presentation and avoid "death by PowerPoint."

- **Chapter 7: Persuasive Photos.** How to take photographs that tell a story.

- **Chapter 8: Edit and Repair Still Images.** How to edit and repair images.

- **Chapter 9: Create Composite Images.** How to create images using compositing techniques.

FOUR FUNDAMENTAL THEMES

- Persuasion is a choice we ask each individual viewer to make.

- To deliver our message, we must first attract and hold a viewer's attention.

- For greatest effect, a persuasive message must contain a cogent story, delivered with emotion, targeted at a specific audience, and end with a clear Call to Action.

- The *Six Priorities* and *the Rule of Thirds* provide guidelines we can use to effectively capture and retain the eye of the viewer.

THE SIX PRIORITIES

These determine where the eye looks first in an image:

1. Movement
2. Focus
3. Difference
4. Brighter
5. Bigger
6. In front

Hausman's laws:

Law 1: Everyone worries about having too little material for a presentation and running short.

Law 2: Nobody *ever* has too little material and runs short.

Law 3: A lot of people giving presentations therefore have way too much they want to cover, they try to cram in and regurgitate too much information, and thus they turn the occasion into a desultory data dump. And they talk too quickly as a result of their panic to wedge everything in.[1]

CARL HAUSMAN

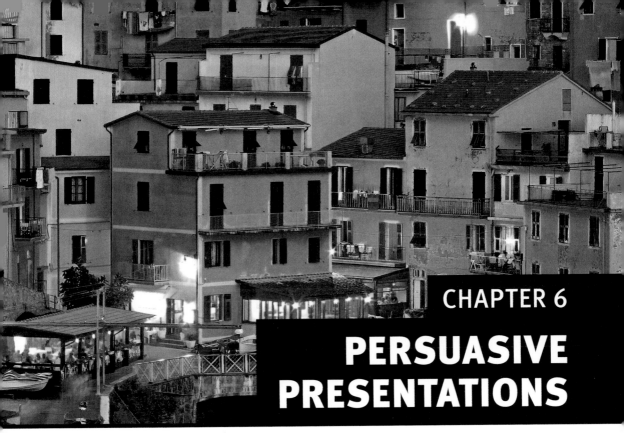

CHAPTER 6

PERSUASIVE PRESENTATIONS

The goals for this chapter are to:

- Explain what is important in a presentation and what isn't
- Help you plan your next presentation
- Create effective slides by integrating what we learned in Section 1
- Present a technique that will help improve your presentation skills

People are not attending your presentation because they want to see your slides. They are attending because they want to see you*!*

OH, I PLEAD *SO* GUILTY to violating Hausman's laws.

I've given almost 500 presentations during my career, maybe more. Before each of them started, I was awash with the feeling that everyone in the audience knew more about my subject than I did. Fortunately, that was not the case, or I would not have been invited to create so many presentations.

But the feeling was there, every time.

I mention this because we have all suffered "Death by PowerPoint," the meaningless deluge of too much data crammed onto a single PowerPoint slide of which the speaker insists on reading every word. We all know that this is really, really bad, and, yet, it is very, very easy to do. My goal in this chapter is help us find ways to make our business presentations less homicidal.

Persuasion vs. Training

I think there is a big difference between a "persuasive" presentation and a "training" presentation. In the first, we want to intrigue and excite the viewer into making a change. In the second, we want to reassure the viewer that they are able to learn and use this new knowledge.

I view a persuasive presentation as fast-paced with high energy. I view a training presentation as energetic, because we still need to keep the attention of the audience; however, a training presentation is slower than a persuasive presentation so that the audience can keep up and learn the material.

I think training presentations can be wordier, slower, and more comforting. Where possible, training presentations should also feature handouts so the viewer can concentrate on listening, not taking notes. Put the details in your handouts, not in your slides.

Still, the focus of this book is persuasion. This is why, before we get deeply into persuasive techniques, I wanted to clarify the differences in style and content between training and persuasion.

Take a Deep Breath

The first thing most of us do when we are tasked with giving a presentation is fire up PowerPoint, Keynote, or any of the online presentation resources, and start designing slides. That way lies failure.

People are not attending your presentation because they want to see your slides. They are attending because they want to see *you*! The fact that you've been invited to give a presentation means that other people recognize you as a subject-matter expert—someone whom they can learn from. Once you accept that you have something that others are interested in learning, this whole presentation business gets a lot easier.

Books have been written on how to give an effective business presentation; I have six of them on my desk as I write this. While each of them makes excellent points, they all boil down to this:

- Think about your audience and what they want to learn.
- Think about an effective way to help them learn it.
- Tell them what you want them to do now that they *have* learned it.

Before we create anything, we need to define our audience and message. Then, this age-old teaching maxim kicks in: "Tell 'em what you're gonna tell 'em. Tell 'em. Then tell 'em what you told 'em." Simple.

It's simple in abstract, perhaps, but tricky in practice.

Think back to Chapter 1, "The Power of Persuasion," where we explored what "being persuasive" means. Persuasion is a dialogue between you and your audience. It's a one-on-one conversation where your goal is to create such a compelling desire in your audience that they want to do what you are suggesting.

Visuals can help, but they can't compensate for a bad, unrehearsed, or unfocused speaker. When businesspeople are asked to list what makes a bad presentation, the list invariably includes the following: lack of preparation; no humor; no visual aids; poor timing; too long; too fast; too slow; mumbled words; no eye contact; inaudible; too loud; confusing; was bluffing (not an expert); sarcastic; poor logic; waving arms; condescending; no theme; nerves; uncomfortable room; didn't answer questions; or unrehearsed.

It's a depressing list! I'm sure each of us has our own list of speaker screw-ups. But, what if we flip those negative comments into positive statements? Suddenly, we have the keys to a successful presentation:

- Preparation
- Delivery
- Visual aids
- Handling and answering questions[2]

[2] (Collins, 1998, p. 16)

The tools don't make the message. We make the message.

Phillip Khan-Panni, cofounder of the Professional Speaking Association, lists seven components of a successful presentation:

1	It is better to demonstrate than to claim.
2	Know how your offering benefits customers.
3	Believe in what you offer.
4	Follow a disciplined approach.
5	Treat a pitch as a joint venture.
6	Engage the emotions of your audience.
7	Show your audience the benefit of what you propose.[3]

Look at how Khan-Panni's seven components echo the themes first expressed by Dale Carnegie: Focus on benefits, not features; understand your audience; and engage the emotions.

I can't help train your voice or assist in answering questions, but I *can* help improve your visuals. And those visuals, while not the star—that's *you*—are still important. Research shows (**FIGURE 6.1**) that we remember:

- 10 percent of what we read
- 20 percent of what we hear
- 30 percent of what we see
- 50 percent of what we see and hear[4]

FIGURE 6.1 How much we remember is based upon how we perceived it.

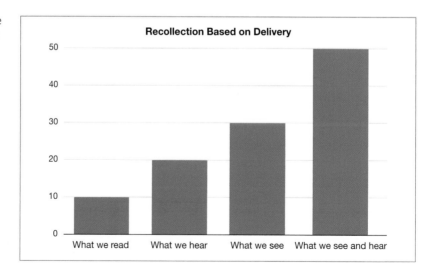

Recollection Based on Delivery

[3] (Khan-Panni, 2012, pp. 8–9)
[4] (Collins, 1998, p. 54)

This means we need to develop visuals that support what we say, rather than create visuals that stand by themselves. Remember, you are the star of this show. Your visuals are the supporting cast.

Planning Your Presentation

Planning your presentation doesn't need to be complex; in fact, any plan is better than no plan. John Collins, writing for the American Management Association, suggests there are six criteria to help us plan any presentation:

- WHO is attending?
- WHAT do they want to hear?
- WHY am I doing this?
- WHEN am I doing this?
- WHERE am I doing this?
- HOW can I put my message across?[5]

Carl Hausman is a professor of journalism at Rowan University and author of *Present Like A Pro*. He writes:

> **A Book to Improve Your Presentations**
>
> As part of my research for this book, I really enjoyed reading *Present Like a Pro* by Carl Hausman. He's a professor of journalism at Rowan University and author of several books on media. His irreverent attitude combined with in-depth advice on giving a presentation make for a fun and worthwhile read.

> Skip the fancy software tricks; so many presentations have become virtual PowerPoint competitions that the genre is almost a joke....You don't always need slides. If you do use them, make sure that:
> - The audience can see your slides.
> - The audience can understand your slides.
> - The slides reinforce your message.[6]

The best place to start, because our presentation is *for* an audience, is to find out what that audience needs to know then offer a solution.

The goal for our slides is readability. We also have an interest in leveraging the fundamentals we've spent the first part of this book learning. So, let's review what John Collins suggests make bad visuals:

- Long lists stating everything the speaker is going to say
- Unnecessary pictures
- Overcomplicated charts, graphs, and diagrams
- Elements too small to read
- Designed to impress the audience, but not support the speaker
- No pictures when one is needed
- Uses only words when an image is needed[7]

[5] (Collins, 1998, pp. 44–45)
[6] (Hausman & Agency, 2017, p. 56)
[7] (Collins, 1998, p. 65)

This is a production still from the film *Hidden Figures*. It reminds me that making our presentations simple does not mean dumbing them down. In this scene, Katherine Goble, working as a human computer for NASA in the early 1960s, is giving a presentation to other engineers. It is technically complex, it is graphical, and it makes sense only to the people in that room. Still, she made her point. You don't need to create a "simple" presentation; you need to create a presentation that is clear and understandable to your audience. (Image Credit: Moviestore Collection Ltd / Alamy Images)

In other words, *you* are telling the story, not your visuals.

If there's one rule to follow, it's this: Keep it simple, stupid! (more commonly called the KISS Principle.) People are attending to learn what you have to share, not to admire your visuals. Develop a clear message, let your visuals reinforce your message, then deliver your message in an engaging manner. Making a presentation isn't hard; it's just that we have so many different tools at our disposal it's easy to forget that the tools don't make the message. *We* make the message.

Backgrounds and Fonts

There is a variety of presentation software available. In these examples, I'm using Microsoft PowerPoint and Apple Keynote. From the perspective of persuasiveness, they both deliver the same results. Pick the tool that feels most comfortable to you.

As you start designing your slides, you need to know how bright the room will be where you are presenting. If the room is dark, a dark background to your slides will make them easier to read. If the room is light, a light background will make your text pop.

To Cloud or Not to Cloud

It is probably the curmudgeon in me, but I prefer apps that store data locally, rather than in the cloud. Also, while it flies in the face of current software marketing, I tend to prefer apps that don't charge a monthly subscription fee. I like being able to access my data without worrying about a monthly payment. You are welcome to disagree.

For example, in a dark room, when the screen suddenly displays an all-white background, your audience starts squinting just to read the text. By making the background darker, their eyes will see your text first. In a lighter room, it may be more effective to have a lighter background so that the screen tends to blend in with the walls, leaving the audience to concentrate on your text, which would be dark against that lighter background.

The two backgrounds in **FIGURE 6.2** are good examples of why you need to consider your room *before* designing your slides. The pages of this book are white, so the white background disappears into the page, and your eye goes right to the text.

With the black background, you see the black box first, because it is different from the page and bigger than the text inside it. Next, your eye goes to the text. The black background is just an extra, and unnecessary, step in getting the eye to see what we want it to see.

Let's stay with a white background for a minute. In **FIGURE 6.3**, which of these two fonts best represents the Keepsakes product? You're right—the serif face, which also gives a feeling of handwriting to make it seem even more personal. The sans serif font feels cold, impersonal, and wrong.

Based on the Six Priorities, the eye goes to that which is bright. But, if the entire screen is bright, there's nothing specific to attract the eye, so it jumps down to the next step: that which is bigger or different—in other words, your text.

FIGURE 6.2 Where does your eye go first? Which is easier to remember: the black text on a white background (left) or a black box with some white text inside (right)?

FIGURE 6.3 Which font most closely represents the feelings associated with these products: Calligraphic 421 BT (left) or Futura (right)?

Change the product, though, and the fonts need to change as well. In **FIGURE 6.4**, which of these two fonts do you want to clean your house? All I changed was the font, but the sans serif font creates a feeling of modern, high-tech quality—everything you demand of the latest in robotic cleaning technology.

FIGURE 6.4 Now, switch products. Which says high-quality, high-tech cleaning robot: Futura (left) or Baskerville (right)?

The three backgrounds in **FIGURE 6.5** feature deeply saturated colors, which overwhelm anything placed on top of them. Yet, these are default color choices in the software. (These backgrounds remind me of a restaurant's background music that is so loud that you can't comfortably chat with your companions.) Backgrounds need to enhance, not obliterate.

Also notice that, with the blue and red backgrounds, I needed to add a drop shadow to the text just to be sure you even noticed it!

Instead, I prefer darker, less saturated gradients or textures, as illustrated in **FIGURE 6.6. FIGURE 6.7** shows the color settings for this steel gray gradient. I moved the Hue selector much closer to the center to remove saturation, while the grayscale slider below was dragged right to darken both the starting and ending colors.

FIGURE 6.5 Notice how the deep saturation of these three backgrounds totally dominates the text. Backgrounds need to support the foreground, not dominate it. (The font is Futura Medium.)

FIGURE 6.6 Darker, less saturated gradients make text easier to read.

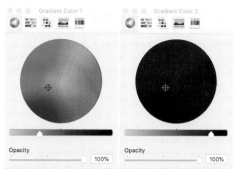

FIGURE 6.7 Here are the color settings for this background gradient, which I modified from one of the default backgrounds.

Secret Tip to Lock the Hue

Here's a secret tip to working with virtually every Mac color wheel. If you press the Shift key while dragging the color "puck," movement is constrained to a straight line between the current location of the puck and the center of the Color Picker. This allows you to maintain the same shade (hue) of color but increase or decrease its saturation. I use this trick all the time.

The background still has visual interest but doesn't conflict with the text, while the text is easy to read. (Also, in these examples, I'm not using white text, but a light gray. It's less garish than pure white but still bright enough to catch the eye at first glance.)

When the room is dark and people are concentrating on the screen, I prefer to use backgrounds that have textures, such as the two examples in **FIGURE 6.8**. Keep the textures dark, with minimal detail, so the text is easily readable. Textures with excessive detail often conflict with serif fonts. So, if you plan to use textures, be sure to use a solid sans serif font.

Textures, to me, provide a more interesting visual background than a solid color. Remember to keep the texture dark enough (or light enough in the case of a bright room) that your text and supporting images are still clear and easy to read.

FIGURE 6.8 Examples of textures; the image on the left has a "stone" look, while the one on the right has a machine feeling. Both backgrounds are dark, with their details obscured.

The background should never call attention to itself.

Notice also that my backgrounds are not perfectly gray; they lean toward slightly warm (gold) or slightly cool (blue). I find a little bit of color makes them less boring than pure gray.

The purpose of a background is to provide an environment for whatever you put in front of it. The background should *never* call attention to itself.

Charts

The reason for using a chart rather than text is that a chart will convey your message more clearly than a written description. However, if you do use charts, explain them. Many presenters put charts on their slides without mentioning them at all, or they breeze over them in a sentence or two, which only confuses the audience as to what they should be looking at.

Instead, if you are highlighting a chart, take the time to explain the graph, its components, and how the data was measured. Then analyze it for your audience. If the chart is important to your presentation, it is important that you explain it during your presentation.

Charts Make Your Points Visually

Take a look at the 3D bar chart in **FIGURE 6.9**. See how quickly it allows you to compare results between months and year; it's much quicker than looking at a table of numbers. Our brains are very good at gaining meaning quickly from visual images.

However, a caution: 3D charts can unintentionally exaggerate certain parts of the graph. For example, if two bars in a graph are exactly the same height, such as the green April and July bars, but the graph is 3D and angled, the one closer to the audience will look bigger, although the bars are the same size. This means an audience can easily misconstrue your data when you use a 3D chart!

FIGURE 6.9 3D bar chart, which illustrates differences (created in Apple Keynote).

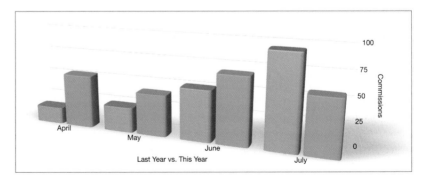

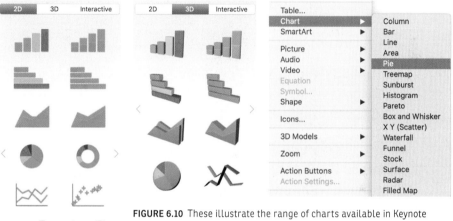

FIGURE 6.10 These illustrate the range of charts available in Keynote (left and center) and PowerPoint (right).

Finally, as you look at this 3D bar chart, note how hard it is to read vertical text. It is even harder when the text goes down, rather than up. Make your text vertical if you *don't* want viewers to read it; make it horizontal if you *do*.

In both PowerPoint and Keynote there are lots and lots and, ah, lots of charts and graphs to choose from; some are even in 3D, as you can see in **FIGURE 6.10**. How do you decide which to use? You choose based upon what you want the chart to show the viewer.

Whether you choose to display elements vertically or horizontally depends upon the number of elements in the chart and how you want the text associated with them to be read. I tend to have small charts display vertically (like the 3D bar chart in Figure 6.9), while charts with lots of elements display better horizontally (like the 2D bar chart in **FIGURE 6.11**).

[8] (Tufte, 1983, pp. 107, 121)

FIGURE 6.11 A PowerPoint
bar chart.

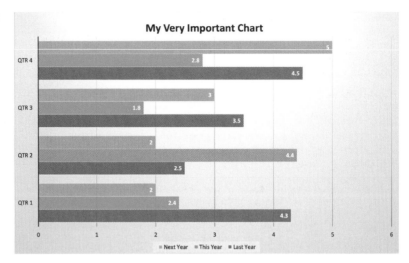

In the PowerPoint bar chart in Figure 6.11, the background is off-white with
a gradient, the text is all horizontal, and the color differences make the bars
easy to read. However, the orange and gray bars have the same grayscale
value, making this a poor choice if it will be displayed in black-and-white.

*Chartjunk can
turn bores into
disasters, but
it can never
rescue a thin
data set.*

—EDWARD TUFTE

Choose the Right Chart for Your Data

Here are some different chart categories and when to use them:

Pie charts. These compare how components contribute to the whole, such as
how different revenue streams contribute to total income. Pies can be displayed open (like a ring) or filled (like a pie).

Bar charts. These compare similar items, for example, savings this quarter
versus last quarter, or Canadian versus US sales.

Column graphs. These show how the same thing, for example, revenue,
varies over time.

Line graphs. These are best when you have lots of data points that change
over time. Line graphs emphasize the shape of the curve connecting the
points more than any individual data point. Line graphs can display lines or
just the points themselves.

Scatterplots. These show how the data clumps, or doesn't clump, into categories. They help us spot associations between data that may not be obvious
by simply looking at the numbers.

And there are even more options! Different chart types can be combined,
sized, colored, and modified until all meaning is lost.

Remember, our goal is readability. Anything that prevents your audience from understanding what your chart is showing is simply getting in the way of your message. A clear message spread across a few charts is much better than a muddled message crammed, with excruciating detail, into one chart.

Just because PowerPoint supports 50 different charts does not mean you need to include *all* of them in your presentation!

Design Thoughts

When it comes to charts and text, less is more! With charts, if you have more than four variables to show, try to split them into two charts; otherwise, the chart gets too busy while each element becomes too small to read.

Limit the number of colors. Keep text black or white if it is against a darker background. Don't color lines or nondata elements, unless necessary for readability. Pick simple colors that aren't too bright or saturated. Try to avoid red, unless conveying a warning. Shades of green and blue tend to work well.

FIGURE 6.12 is an example of "I can, therefore I will." (Designers call this look a "train wreck.") The background is too bright and too saturated and totally dominates the image. The colors seem to be picked by throwing darts at a color wheel. Each line of fonts is different—and none suggests any kind of coherent message. There are at least three emojis more than needed, if this even needs emojis.

Drop shadows create the illusion of depth; varying the size and softness of the drop shadow implies variations of depth. Notice that the size of the drop shadow under the blue line indicates it is some distance from the background, while the title's smaller drop shadow indicates it is close to the background. But, the blue line is *behind* the title. There's a depth mismatch here. Also, the angle of the line's drop shadow is straight down, while the angle of the drop shadows under the text is down right. Lights don't change position like that.

You get the idea—the person who designed this slide forgot that the top two goals in any slide are readability and support for the speaker.

They *did* align the big clown using the Rule of Thirds, though. Yay?

A clear message spread across a few charts is much better than a muddled message crammed, with excruciating detail, into one chart.

FIGURE 6.12 Sigh…Just looking at this hurts.

Working with Images

It seems self-evident to say that slide shows are designed for images, as shown in **FIGURE 6.13**. However, all too often we forget this and bury our slides in text. Images are also good to use as backgrounds, as long as they are not too detailed and don't interfere with the text. If you use an image for a background, make sure it relates to the subject of your presentation.

FIGURE 6.13 The image is not centered. Rather, it follows the Rule of Thirds, allowing you to see the image first then explore within the frame.

There are many things going on in this example:

- The image isn't centered; instead, it is placed according to the Rule of Thirds. This allows your eye to wander but always come back to the image.
- The image is in focus, big, and different. Your eye goes there first.
- The title is brighter and follows the line of the river; your eye goes there second.
- The image isn't centered, so your eye then starts exploring around the frame and discovers the subtitle in the upper-left corner.
- The background image of clouds reinforces the theme of standing outdoors under a big summer sky, yet without a lot of detail to distract from the foreground.
- I added a blur to the background image in Photoshop then decreased the color saturation a lot so that it became a steel gray sky without darkening the clouds, which would have implied rain.
- Plus, there's a nice pun in the title because, after all, you want the audience to have fun.

Everything you create should reinforce the message you want the audience to remember. As you create your own presentations, remember that you are not locked into using solid-color backgrounds. Take advantage of the Rule of Thirds, the Six Priorities, and the other visual techniques we covered in Chapter 2, "Persuasive Visuals," to liven your presentations.

Also, notice that I didn't put a lot of words on the slide in Figure 6.13. In the next slide for this presentation, I would discuss the "what" and "when." Then, on the last slide, I would present a Call to Action specifically telling the audience where to go to watch.

More Thoughts on Drop Shadows

We use drop shadows for one reason: to separate an element from the background. This could be to make text more readable or to imply distance between the foreground and background elements.

In **FIGURE 6.14** the text has the same color as the background, something I recommend you try really hard not to do, though white text against a bright sky is a pretty common problem. Without the drop shadow, the text would be unreadable. But, with it, the title is readable, especially because the background texture is so plain, it doesn't compete with the text.

The two images are positioned according to the Rule of Thirds, with the larger image in the background. We know this for two reasons: the image of the duck is in front of the bench, and the drop shadow of the duck is larger and softer than that of the bench.

As you build presentations, look carefully at what you are creating so that all your elements work together coherently.

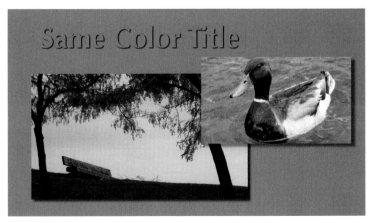

FIGURE 6.14 Three different examples of using drop shadows, though I *strongly* recommend you never use the same color for both title and background!!

Both the position of each image and its drop shadow reinforce the foreground position. As images get farther from the background, the drop shadow needs to get softer (blurrier) and less opaque. The bottom image has an opacity of 62 percent, with an offset of 15 pixels. The top image (the duck) has an opacity of 35 percent with an offset of 25. (Text shadows needs to be darker; I generally give the drop shadow for text an opacity setting of 95 percent.)

As you build presentations, look carefully at what you are creating so that all your elements work coherently together.

A Cool Design Tip

Carl Hausman made an interesting point that I hadn't considered before: "I'm a fan of numbers in slides…. For whatever reason, a number coupled with a good visual design is arresting and commands attention…. To further reinforce my point, do an image search for 'great PowerPoint slide examples' and note how the ones that jump out at you often feature a number and an image. When there is text, it is generally short and to the point."[9]

> *I'm a fan of numbers in slides.*
> —CARL HAUSMAN

FIGURE 6.15 Keep your slides simple, the text large, and don't hesitate to overlap objects.

You can see an example of this in **FIGURE 6.15**. The first question the viewer asks is, "50 percent of what?" They have to pay attention to you to get the answer. Again, this throws the spotlight back on you, which is where it needs to be. Also, notice that by overlapping "50%" with the image, we are able to make both bigger. There's no rule that says each element needs to be in its own little box.

Transitions: Less Is More. Really.

When I teach video editing to high-school students, they can't wait to get to the section on transitions (**FIGURE 6.16**) and animation (**FIGURE 6.17**). Once we get there, I know I'm going to lose control of the class. Students jump out of their chairs to show their latest creations to their friends; a lot of giggling ensues. Effects are exciting, cool, and fun to play with, especially if you are the one creating them. For the rest of us, though, if I see one more Page Curl, I'm going to throw my shoe at the screen.

Let's take a step back. People are not attending your presentation to see your slides. If they were, cancel the meeting and email a PDF. Save everyone time. They are coming to see you. As soon as the slides start to upstage the

[9] (Hausman & Agency, 2017, pp. 56–57)

FIGURE 6.16 PowerPoint has more than 40 transitions, each of which can be modified in multiple ways.

FIGURE 6.17 The Animations panel in PowerPoint.

speaker, you lose control of the audience. Never intentionally lose control of your audience; you'll never get them back.

For this reason, consider each fancy transition between slides as a blaring horn. The first time it's eye-catching. The second time, it's okay. After that, though, people start eyeing the exits.

There is one exception to this. I do use transitions to build text out on a slide. Rather than reveal the entire slide at one time, I'll build each bullet point individually for key slides. This allows me to focus the attention of the audience on the point I'm making at that moment. I also often use humor in my talks. By not showing all my text at once, I can build to a punchline without my slides giving it away.

Still, like transitions between slides, the more complex the text build transition, the faster it will grow old. Keep things simple and you'll be able to use the effect more.

Keep things simple—leave effects, like sounds and transitions, for the high school kids to play with.

Media. Gently, Please.

Everything I just wrote about transitions goes doubly for sound effects. Like humor, it can easily go astray. What sounds funny to you may not sound funny to anyone else. Worse, putting a sound effect with each slide gets tiring really, really fast. It's like a silly ringtone that loses its humor by the third ring. My suggestion is not to use sound effects, or, best case, use only one per presentation, say a fanfare when you announce the new thing or a thud when the new low price slides down. Frankly, though, I suggest you leave effects, like sounds and transitions, for the high-school kids to play with.

When dealing with media, also consider what happens if the sound system or projector dies in the middle of your presentation. I've had both happen.

It all comes back to you being the star. As soon as you delegate "stardom" to your slides, you've lost control.

That doesn't mean you shouldn't use media. You can and should, but in limited bursts. When someone is on stage, live, the audience is leaning forward, actively listening to what's being said. As soon as you play a clip, your audience shifts back into passive listening mode, letting the content wash over them, as if they were sitting on their couch at home. Once the clip is over, it takes a lot of extra energy from you to reengage them back into active listeners.

KEY POINTS

PowerPoint and Keynote are powerful presentation packages with lots of features, effects, and capabilities. In the hands of a design expert, they can do amazing things.

But, for most of us, that's more capability than we need. We are at our persuasive best when *we* are the ones doing the talking, not our slides. A presentation doesn't need to be technically complex to be effective. No one wants to watch a presentation where the presenter isn't interested in presenting. In those situations, show a video.

A live presentation puts you front and center, delivering a focused message, targeted at a specific audience and supported by simple, clean slides that use text and images to reinforce, not repeat, what you have to say. Remember:

- You are the star of any presentation, not your slides.
- Focus your message and target it at your audience.
- Show, don't just describe.
- Reinforce key points with visuals.
- Emphasize readability in your visuals.
- Reinforce your message through the design of your slides.
- Use charts only when text is not sufficient.
- Use images when it is easier to show than to explain.
- Never let your slides upstage you.
- Your presentation is a conversation between you and the audience, *supported* by your slides.

It all comes back to you being the star. As soon as you delegate "stardom" to your slides, you've lost control.

PRACTICE PERSUASION

A Pecha Kucha talk (which means "chit chat" in Japanese) is a 20-slide presentation where each slide automatically changes after exactly 20 seconds.

It was invented in 2003 by two architects looking for ways to streamline long design presentations. It quickly morphed into a series of nighttime get-togethers. Since then, Pecha Kucha talks have spawned an entire cottage industry where technology meets storytelling.

Pick a subject you feel passionate about. Create 20 slides to illustrate it, set them to automatically change every 20 seconds, then present the talk to your family, your cat, or the wall. Learn more at www.pechakucha.com.

- Create slides that are image-focused. The less text the better.
- Focus your message to fit into short, bite-sized chunks.
- Practice telling a story, comfortably, in a short period of time.

PERSUASION P-O-V

EMPOWERING CHANGE

Ron Melmon
Producer, Zippo Productions (zippoproductions.com)

Our company specializes in creating documentaries, and over the years we've learned about the power of film to change minds.

I was reminded of this recently as we were working on an inspirational doc about Telluride, Colorado, called *The Valley*. The film focuses on the beauty of the area and what residents are doing to protect it.

As we were editing the film, prior to its premiere at the Sedona International Film Festival, we were contacted by a group of citizens in Sedona who faced a similar problem: a developer wanting unrealistic zoning changes.

About 15 people from a Keep Sedona Beautiful group came over to our house for a sneak preview of the movie. The story of *The Valley*, and the success of Telluride citizens in maintaining their environment, motivated the people from Sedona to get organized and focus their efforts. The townspeople were so effective that the developer withdrew their application for a zone change.

I've seen a lot of films move an audience. This was the first time one of my films was used as a teaching aid to create political change.

A live presentation puts you front and center, delivering a focused message, targeted at a specific audience and supported by simple, clean slides that reinforce, not repeat, your message.

Photographs, more than almost anything else, have a special emotional appearance: they are personal, they tell stories.[1]

DONALD A. NORMAN
DIRECTOR, THE DESIGN LAB
UNIVERSITY OF CALIFORNIA, SAN DIEGO

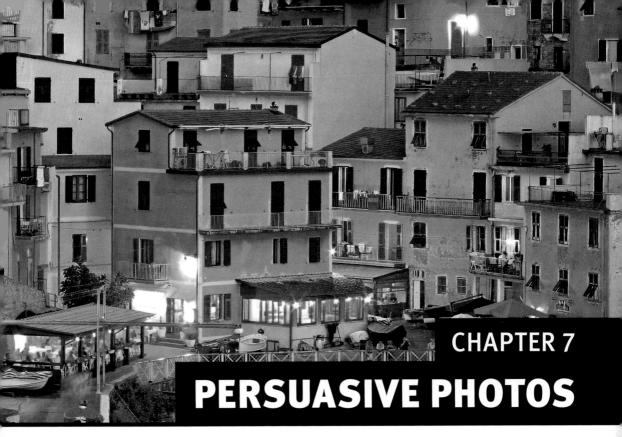

CHAPTER 7

PERSUASIVE PHOTOS

Building on concepts first introduced in Chapter 2, "Persuasive Visuals," the purpose of this chapter is to illustrate simple techniques that will improve the quality and impact of your still images. Specifically, this chapter will:

- Explain the difference between using stock or original images
- Help you plan a photo shoot
- Illustrate the basics of lighting
- Discuss how to work with actors/talent
- Illustrate different composition techniques

IF YOU WANT TO FEEL INTIMIDATED, go to the library and look at all the books about how to take photographs. They fill *shelves*! The sheer quantity and diversity are overwhelming. How can we cover all this in a chapter?

Here's the short answer: We can't. If you want professional-grade images, hire a professional to take them. This is the same as any other craft skill— experience, equipment, and talent make a difference. In fact, most of the photographs in this book were taken by pros.

But, sometimes, that isn't possible. You may need a quick image for social media, there's no budget for a pro, or you want to test some ideas. That's where this chapter can help. My goal is to show you different ways of telling stories and evoking emotions using photographs. Once you see the differences, you can pick the techniques that work best for your projects.

Here are the basics of how to improve the quality, impact, and storytelling of your images.

Horizontal or Vertical—Which to Pick?

Bar fights have broken out over whether it is better to shoot images as horizontal (left) or vertical (right). What's the answer? Um, both. The decision on what to shoot is determined by what you need to deliver. If you are shooting for vertical ad displays in a mall, you'd be foolish to shoot horizontally. If you are creating images for video, horizontal is the way to go. For this book, I'm mostly using horizontal images because they fit the book format better. Let your deliverables decide! (And, if you need to deliver both horizontal and vertical images, plan to shoot each orientation separately.)

Where Do You Start?

Before you hire a photographer or start looking for images, take a step back and think about what you want to accomplish. (Yes, I know, that's the boring part.) Remember Ben Franklin's adage: "Those who fail to plan, plan to fail." Our job, at least in terms of this book, is persuasion. But persuasion is more than someone pretty smiling at the camera; smiling presenters are, all too often, overused. Our primary goal is to attract the eye. There are many ways to create compelling images that don't involve beautiful models and wide smiles. It's far better to hook the viewer's imagination. "Slice of life" is a term used to describe images that represent "the real world," in hopes that viewers identify with it and continue watching.

Those who fail to plan, plan to fail. —BEN FRANKLIN

So, in planning the images you need, first answer these questions:

- What is the message?
- Who is the audience?
- What is the story we want to tell?
- What do we want the audience to do?
- What kind of image is most likely to get their attention?

These answers don't always need to be detailed, though bigger-budget projects will benefit from scripts, storyboards, budgets, and schedules. But, at a minimum, take time to write down your answers. The process of writing will force you to think through what you are doing before you start.

"The creative process" is both fun and frustrating. I have always enjoyed creating stuff, from books to television programs. But, I've also learned that there are so many creative options that if I don't take the time to plan, I'll find myself in the middle of a project with no clear idea of where I'm going or how to know when I'm done.

Our goal is persuasion, and we are putting our time and money behind what we create in order to attract an audience and effect change. Time spent planning will return dividends on your investment.

After your plan is complete and you know what you want to do, you come to your first big question: Where do you go for images? You have two basic choices: Obtain them from a stock house or shoot them yourself. (You can also steal them from the web, but do you really want you, your company, or your clients to run that legal risk?)

Images used for education or editorial coverage are a special case, called Fair Use. The rules are different, but copyright still applies.

A Caution on Copyright

All creative works are copyrighted at the moment of creation. While copyright laws vary by country, in general, the person creating something, for example, a photographer—whether paid or volunteer, amateur, or pro—owns all rights to the image. While copyright is created automatically, it is a good idea to "register" a creative work, where possible, to protect your rights in the event of an infringement. Understanding and honoring copyright is important, because the legal damages for infringement can be severe for you and your clients.

If you are hiring someone to create something—text, images, video—the creator of the work owns the copyright. Period. The owner of the copyright decides who gets to use their work. This means that, as part of their contract, there needs to be clear language on transferring rights from the creator to you.

If the creator is an employee, they still own their creative work, unless there's explicit language in their employment contract stating who owns creative work created on company time for the company.

If you want to reproduce, publish, or sell an image or include an image as part of a larger composite image, you need to have the legal rights to do so. Obtaining these rights is called "getting permission" or "licensing." Sometimes those rights are free; other times you'll need to pay for permission to use an image. Obtaining permission does *not* transfer ownership; it just means that you have a license (permission) to use that image for a specific purpose and time.

Images posted to the web are also covered by copyright. Posting an image does not mean it is free to all users. While the web is more *laissez-faire*, copyright still applies. Failing to obtain proper rights can land you in a lot of legal trouble. The fines are significant, and "I didn't know" is not a defense. Now you know. Copyright is complex and favors the creator. When in doubt, consult a lawyer.

A stock image/footage house is a company that licenses images, audio, or video clips for use by content creators. I'll refer to these companies as "stock houses."

There are many reputable stock houses but far more dodgy ones, so it's always good to check references before basing a campaign on their work. Sites that I've used and like include: Pexels.com, Pond5.com, Shutterstock.com, Getty Images, Adobe Stock, and others. A Google search will turn up links and reviews.

There are several big benefits to using stock images (or footage):

- The image already exists, so you can see exactly what you are getting.
- There are millions of images to choose from.
- Pricing is reasonable, ranging from free to a few hundred dollars for special or rare images.

The big disadvantage to stock is that you may not be able to get *precisely* what you need. In many cases, that's not a huge limitation. But, if you need to feature a new product or illustrate something specific (like the images on lighting later in this chapter), you will get better results hiring a professional photographer to create exactly the image you have in mind.

Let's talk about pricing for a minute. When you license a photo—remember, licensing does not grant ownership, just the permission to use it—it is up to the owner of the copyright (or their agent, the stock house) to decide whether they want to charge for the image and, if so, how much. Many photographers earn their living through the money they make licensing their images.

Other sites offer "free" images. Some of these sites, like Pexels.com, are reputable and offer image licenses at no charge. Others, though, steal photos from other sites then offer it for "free," without providing the necessary licenses. If you are building a campaign, be sure you have the rights. If you don't, maybe you won't get caught. But, if you *do* get caught, the results won't be pretty. This is why people use reputable stock houses and pay for their images. For just a few dollars, this potential copyright liability goes away.

People use quality stock houses and pay for their images to make sure any potential copyright liability goes away.

Plan Your Photo Shoot

Let's say you need to create original images for your next project. The first step is *not* hiring a photographer. Rather, take your plan and turn it into a list of the specific images that you need.

I made this mistake when writing this book. Once I realized that I would need original photography for this chapter, I got really excited and called a couple of photographers to find out how much it would cost. They asked:

- What's your budget?
- How many shots do you want?
- Where do you want them shot?
- Do you have permits for where you want to shoot?
- How many actors do you need?
- Do you need props, costumes, hair, and makeup?
- What media format do you need?
- Do you need retouching?
- And what's your deadline?

Plan Your Locations

As you plan your shoot, also think about *where* you want to shoot. If you don't own the property, you may need to get a permit to shoot there. Amateur photography is allowed in lots of places. However, if you plan to make money from your work, called "commercial photography," you may need to get written permission to shoot there. Many times, permission is free and provided via an email. For other places, you may need to fill out a form and pay a fee. Again, it never hurts to ask, and doing so can save legal problems down the road.

Throughout this book, I've said we need to plan. Well, in my enthusiasm, I forgot to take my own advice. A good photographer is not going to simply point their camera and start clicking. Rather, like a film director, they want a "script" that explains the story you want to tell and the specific images you need to tell it. The more detailed, the better. They will then apply their skills and creativity to your "script" to make these images as good as they can.

So, before I could pick the images, I first had to write this chapter; that told me which images I needed. I next compiled a detailed shot list and hired cast and crew, then we started shooting (**FIGURE 7.1**).

FIGURE 7.1 Our photo team in action, from left to right: Kim Acuña, Allison Williams (actors), Janet Barnett (photographer), and Amy Camacho (hair and makeup)

It All Starts with Light

Photography is all about the manipulation of light falling on a subject to create a pleasing image. To achieve this result, we need to determine *what* we want to shoot (the content) and *how* we want to shoot it (the technique). There are three main techniques I will cover in this chapter:

- Lighting
- The position of our actors (called "blocking")
- The position and framing of the camera (called "composition")

We first looked at camera position and composition in Chapter 2. Now, I want to go deeper into blocking, composition, and, most importantly, lighting.

Lighting is magical. It converts the ordinary into the extraordinary. It transports, reveals, and highlights. Or, it looks like a snapshot taken by your nephew.

Actors vs. Talent

When I'm referring to the folks who appear on camera, I generally call them "talent." Not all of us are actors, which I consider a special breed of talent. Most folks on camera today are "presenters." Appearing comfortable on-camera, communicating well and gracefully, and having the audience enjoy spending time with you requires a special talent. So, for the rest of this book, I'll refer to people who appear on camera, or on mic, as our "talent."

When we move into the studio, we have complete control over lighting, blocking, and composition. We want to make our talent look natural and comfortable, without harsh shadows or squinting. To do this, lighting directors for film and video invented a technique a long time ago called "three-point lighting." This is a system that actually uses more than three lights.

The main light, as you see in **FIGURE 7.2**, is called the "key." The key provides both the dominant light and the dominant shadows. The key is dominant, but it is not in front; it's slightly to the side. The background is lit with the "set" lights, and the talent is further defined using a "fill" light, which "fills in" the shadows created by the key, making them less intense. The "backlight," or hair light, helps separate the head and shoulders from the background, providing a more three-dimensional look.

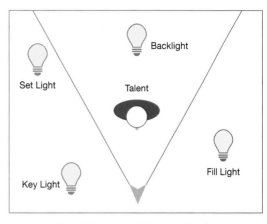

FIGURE 7.2 Typical three-point lighting with talent in the middle and a set in the background. The gray triangle represents the position and view of the camera.

That phrase "dominant shadows" is important. That which gives our faces—or anything, really—shape and dimension are its shadows and texture. We don't want to make shadows disappear; that tends to be unflattering. We want to use shadows, highlights, and differences in light levels to shape the subject in a way that makes it look real, dimensional, and separate from the background. We call this shaping process "modeling." Three-point lighting provides this.

Not All Lighting Is the Same

I've spent my life on film and video sets, where we use lights that are always on, tend to create sharp shadows, and are placed using variations of three-point lighting. But, this is not the typical technique for still photo lighting. There, giant soft boxes, driven by strobes to over-power the ambient light, are much more common. For the photos in this chapter, the photographer and I worked together to create a lighting style that we were both comfortable with. As with any creative project, collaboration is key, as is making sure everyone is speaking the same language.

I should also mention that for every rule, there's an equally good reason for breaking it. I'm not saying that this is the only way to light. But, if you tend to shoot using overhead ceiling lights, these techniques will make your images look a *lot* better!

We want to use shadows, highlights, and differences in light levels to shape the subject in a way that makes it look real, dimensional, and separate from the background.

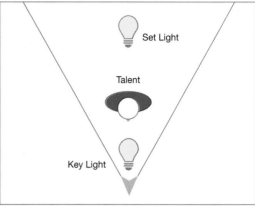

FIGURE 7.3 The light is coming from directly in front of the camera. The sun provides a strong backlight, but her face is very flat; there are no shadows to provide modeling.

For example, **FIGURE 7.3** illustrates the problem with light coming directly from the front: There are no shadows. There's no depth, no modeling. A face is defined by shadows and textures. Front light washes them away. This might be useful if you are trying to make someone who is 70 look like they are 25. But, more generally, the image just looks flat, as Kim does in Figure 7.3.

This is one of the reasons I don't like the ring lights that fit over the lens of a camera. From the camera operator's point of view, they are easy to use. But, as someone who has had to stare into those things, they force me to squint and put a barrier between me and the lens. These lights also make talking to the camera that much more intimidating and, worst, remove the lovely modeling that shining a light from the side provides.

A "set" light is any light that illuminates the background. In these lighting examples, the set lights are the two lamps in the background and the wall behind her. Set lights are controlled so that they light the background, not the talent. The key, fill, and backlights are controlled so they light the talent and not the set.

FIGURE 7.4 illustrates how we use three-point lighting to provide shape to the talent and depth to the background. This is the lighting you see everywhere in film and TV. While the background is lit using a multitude of lights, the talent is illuminated using a key, fill, and backlight. Notice how this lighting makes it easy to see her face, avoids harsh shadows, and provides separation between the foreground and background. In terms of the Six Priorities, she is in focus and brighter and bigger than any other element in the frame.

Let's deconstruct this lighting to see how it is put together.

In **FIGURE 7.5**, we turned off the key and fill. The image becomes mysterious, almost frightening. We see her hair, caught in some stray light, silhouetted against the background, but we can only imagine what she looks like. For

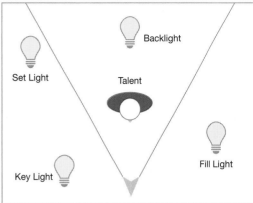

FIGURE 7.4 Here's a lovely example of all lights in use.

silhouettes to work, the background needs to be lit enough to see that there is someone in front of it. As a note, I kept the background dark for these images to make it easier to illustrate the effect of the different lights.)

Control the Drama

A scene is described as "more dramatic" when there is a bigger difference between the amount of key and fill. Allison's image, in Figure 7.6, is dramatic in that the fill is much less than the key. When we move outside in a few pages, the difference between key and fill is negligible. Interview shows use almost equal amounts of key and fill. Dramas use as little fill as they can get away with.

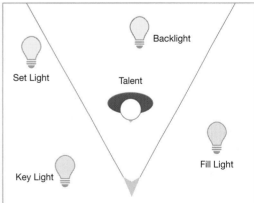

FIGURE 7.5 Silhouette. The back and set lights are on. We see she is there, but not who she is.

In **FIGURE 7.6**, the backlight provides dimensionality to the head, separating her from the background. But with the set lights turned off, we don't know where she is. Also notice how dark the camera-right side of her face is.

In **FIGURE 7.7**, when the background is dark, see how turning off the backlight makes her seem two-dimensional? There's no depth; she merges into the background. If the background is brightly lit, you can get away without a backlight, but, a backlight always helps separate talent from the background.

I call **FIGURE 7.8** the "NCIS-look." It is very dramatic, with strong shadows and modeling. You'll see this look—featuring limited fill—on most current crime and dramatic shows.

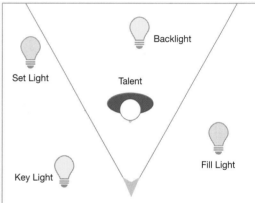

FIGURE 7.6 The set (background) is not lit. Key and backlights are on. Allison is isolated in a pool of light.

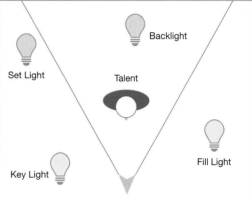

FIGURE 7.7 The key and fill are turned on, but the back and set lights are off.

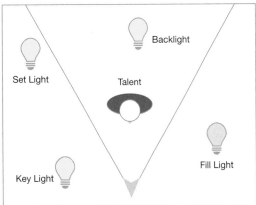

FIGURE 7.8 This uses a strong key, with back and set, but no fill providing a very dramatic look.

FIGURE 7.9 is the final version using all three talent lights and the set lights illuminating the background. Most news, interview, and talk shows use this style of lighting, though with a background that is more brightly lit and a brighter fill light.

Lighting Angles

In general, the key light is 45° to one side of the talent's nose and 45° up. The fill is 30° over on the other side, and 0° to 30° up. The backlight is immediately behind the talent and 70° up. Vary the intensity of your lights to increase or decrease the drama. Consider these positions as starting points then move your lights to get the look you want.

FIGURE 7.9 All lights lit.

These examples illustrate typical studio lighting. But, perhaps you want to make a point by creating a more intense look. Here are three special lighting techniques, each providing an unusual or unsettling effect.

Virtually every office uses ceiling lights. The problem is that they make the people lit with them look nearly dead. Look at the deep shadows under Kim's eyes, under her nose, and under her chin in FIGURE 7.10. It's not a pleasant look at all.

FIGURE 7.10 Top lighting against a dark background. This is just frightening!

FIGURE 7.11 Here the light is under her chin. Very evil.

FIGURE 7.12 Rim light, where the key is moved slightly behind the talent, heightens the emotion of a scene.

If you are trying to make someone look unattractive, as we are doing here, this is a good way to do it. But, if you are shooting your CEO, your job security may depend upon moving the lights to a softer angle.

Be sensitive to your lighting. It tells a story, just like words, color, or the expressions of your talent.

FIGURE 7.11 is the "resident evil" look. The light is placed underneath her chin and turns normal shadows into highlights. Compare lighting from the top with lighting from underneath. Highlights and shadows are reversed. We are so used to seeing light come from the top, such as room light or the sun, that this look, made famous by Bela Lugosi in the film "Count Dracula," is used in every horror film since to signify the evil villain.

As you can see from these examples, there are *lots* of ways to light! I love the dramatic, intriguing look in FIGURE 7.12, where we just barely see the face, yet the head is rimmed in light. This look, too, is a common dramatic lighting technique. (A similar example is on the front cover of this book.)

We can get even more creative by adding colors. For example, we added red to the fill light of Kim in the rim light shot. Or we can use lighting "specials" that highlight just the eyes, shoot in dimly lit dive bars lit with neon lights, or evoke a stroll along the river on a moonlit night. But, those shots tend to raise the budget, and for most of us, we just want to make our talent look good.

Still, there is one more lighting element we need to consider: the sun.

Dealing with the Sun

The sun is bright—really, really, *really* bright. There isn't a light on the planet that begins to equal the light output of the sun. So, once we move outdoors, we need to work *with* the sun, not against it.

Colors will look most vibrant on a sunny day, but your talent will look better under a light overcast, because the clouds soften the sunlight and, thus, the shadows. (High-end film shoots often put a huge translucent sheet, called a "diffuser," between the sun and the talent to soften shadows.)

FIGURE 7.13 illustrates the problem: The sun is directly in front of Kim. She is squinting, which doesn't improve her expression. We need to be careful with these shots; the sun is *so* bright, it is hard to avoid overexposing something or stressing the talent.

In **FIGURE 7.14** we repositioned Kim so the sun is behind her. This makes the background look better, and she is now no longer squinting, But, her face is too dark.

Keep the sun behind the talent so it acts as a "super-back-light." Then, use a white cardboard reflector to balance and reflect light back into their face. Reflect the light slightly from the side, not front, for best modeling and so they don't squint. The image on the right of **FIGURE 7.15** shows our setup for this shot.

FIGURE 7.13 Talent lit in direct sunlight—and squinting.

FIGURE 7.14 Sun behind talent, excessive backlight.

Three-Point Lighting Also Applies to Products

This three-point style of lighting works equally well for products as people. Backlights separate the foreground from the background. Key lights provide dominant light and shadow, while the fill light softens the edges and keeps the image from becoming too harsh.

FIGURE 7.15 This is a low-budget way to deal with the sun. Place the sun as backlight then use a reflector as the key light from the side.

Essentially, when you move outside, use the sun as a backlight, a reflector as the key, and the ambient light as the fill. This is still three-point lighting, but without using lights!

Be sensitive to your lighting. It tells a story, just like words or colors or the expressions of your talent.

Blocking Talent

BLOCKING The positioning of talent and cameras so that the audience sees what you want them to see, when you want them to see it.

Blocking is the process of planning the positioning of talent and cameras, along with their movements, so that the audience sees what you want them to see when you want them to see it. In stills, this is done using frozen moments in time. In video, blocking more resembles a dance.

I want to stress that being on camera is intimidating, even for people who make a living at it. Reassurance and support go a long way to getting great performances from your talent. There's something about staring into a lens that is very discomfiting. As you work with your talent—especially amateurs—do everything you can to reassure them that they are doing well and looking good even if they aren't, because yelling only makes it worse.

> ## Dealing with Inadequate Talent
>
> I learned this lesson about working with inadequate talent years ago at a music recording session. We had a trumpeter who just could not hit the notes. I told the music contractor that we should fire him. Instead, the music contractor suggested that, since we had already paid him, we should let him finish the session. Yelling at him would not improve his musicianship nor save us money. But it would embarrass him in front of peers and make him angry with us. Since we would need to re-record his part anyway, we let him finish then hired a different trumpeter for the next session.

Reassurance and support go a long way to getting great performances from your talent.

When it comes to blocking, I want to expand on a subject that applies to stills and has an even more important role in video: the 180° Rule.

The 180° Rule says that if you create a line between two people having a conversation, the best wide shots are around the middle of the line, while close-ups improve as a camera moves closer to the line and over a speaker's shoulder. Let me illustrate.

As we learned in Chapter 2, wide shots, such as **FIGURE 7.16**, show the "geography" of a space, for example, where people are placed in relationship to the background and each other. But we don't see anyone's face clearly because we are too wide and our cameras are in the wrong spot for close-ups.

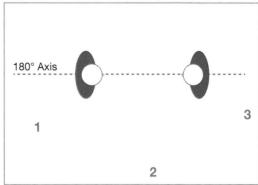

FIGURE 7.16 Kim and Allison are having a conversation. This is a wide shot from the camera 2 position, which is centered between them. The floor plan numbers indicate camera positions.

When students are in my class, I ask them to draw where they would put a camera to shoot someone's close-up. Invariably, they put the camera in the center, at the camera 2 position, which results in the less-than-useful close-up of Kim's ear (FIGURE 7.17). The camera 2 position is good for wide shots but creates profiles for close-ups. This is what I mean by a camera being in the "wrong spot." It looks good on a storyboard, but in real-life not so much.

Remember the 180° Rule. The closer you move cameras to the line *and* behind the shoulder of a talent, the better your close-ups will be. As an example, see FIGURE 7.18. This close-up is from the camera 3 position.

FIGURE 7.17 A close-up from the camera 2 position only shows us Kim's ear.

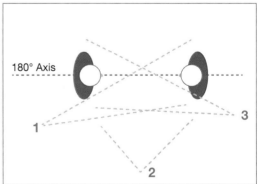

FIGURE 7.18 Move the camera closer to the line and over a talent's shoulder for the best close-ups, for example, positions 1 and 3.

I learned my greatest lesson about working with talent in 1978. I was directing a pilot for PBS on art history and the cathedrals of Europe. After much effort, the producers were able to get noted film actor Vincent Price to star. Vincent, in addition to a successful career in horror films (and Michael Jackson's *Thriller*), was an accomplished art historian and chef. He was intrigued by the project and graciously agreed to host.

As with any production, everyone was nervous as the first day of shooting began. I'd thought a lot about our first shot, which would open the series. It started with a close-up of Vincent's hand running along a line of books. He grabs a book and turns out to the camera as we dolly back from a close-up to a wide shot to discover him in an ornate and beautiful library.

In my mind, it was a glorious way to open the show.

When it came time to shoot, we rehearsed the shot and, then, rolled tape. As Vincent's hand moved past the books, it looked white and wasted. The camera move was shaky, and the final framing was off. Without thinking, I said, "That looks terrible." Vincent took

one look at me and left the set, with the producers in panicked pursuit.

Five minutes later they came back and said, "Vincent feels that you've insulted him. You need to apologize." I felt terrible, because that wasn't my intent.

I went back to his dressing room, apologized deeply, and told him that my reaction was due to the quality of the camera work and lighting, not his presentation.

At that point, he taught me the lesson that I'm sharing with you: Whenever you are in production, whether for stills or video, as soon as you say "Cut," direct your first remarks to the talent. When you are on camera, you feel exposed. The first thing actors want to know is how they did. After you reassure your actors, then talk to the crew. When I said the shot was terrible, I was thinking of the shot, not the people in it. Vincent's performance was fine; where I needed to spend time was refining the camera work.

I never forgot that lesson. Always let your talent know how they are doing *first*; then deal with any production problems.

Always let your talent know how they are doing first; then deal with production problems.

By using the 180° Rule, we can easily determine the best place to put a camera for the shots we need. For example, by moving the camera to the side, as in Figure 7.18, we can easily see faces. We'll come back to this again in Chapter 13, "Video Production," when we talk about shooting dialog. For now, use the rule to figure out the best place to put your camera for each shot.

Framing and Composition

Once your actors are in position, framing determines the "composition" of the shot. The principle visual cue for emotion is the face. Since we are using emotions to convey our content, we want to be sure to emphasize faces as much as we can.

Not only is this done in how we position the camera, as we learned in Chapter 2, but it is also done in how we position and frame the talent. Chapter 2 covered the basics of composition, position, and framing. Now let's take it further.

FIGURE 7.19 The Rule of Thirds divides the screen into thirds. Ideally, put talent at intersections or, at a minimum, on a line. (Image Credit: Mentat-dgt / pexels.com)

FIGURE 7.20 The left image centers her head in the frame; the right image is framed according to the Rule of Thirds. The right image looks much better.

In Chapter 2, I introduced the concept of the Rule of Thirds (**FIGURE 7.19**). This divides the screen into equal thirds both horizontally and vertically. You'll see this framing rule applied to almost all the images in this chapter.

Many inexperienced photographers put the main image in the center of the frame, as illustrated in **FIGURE 7.20**. The problem is that this doesn't create a very pleasing shot. Compare the image on the left, where Kim's head is centered, versus the shot on the right, where both the wine glass and Kim are framed according to the rule. See how much more interesting and involving the shot becomes? By following this rule, you'll also decrease the amount of headroom over the top of the actor's head.

FIGURE 7.21 provides another example: holding a prop or product. Psychologically, we want to hold it away from our body. But when we do, the shot becomes too wide. We can't see the product or the presenter clearly. Instead, pull the product uncomfortably close to your face. This allows for a tighter shot, enlarging both the face of the presenter and the label of the product. Emotions live in the face, and we tend to trust people who we can look

FIGURE 7.21 Which has more impact, a wider shot or one that's closer?

directly in the eye. Plus, going closer on the product makes it easier to read and remember its name.

Let's return to Kim at the bar. In the left image of **FIGURE 7.22**, we are eavesdropping on someone's problem. In the right image, we become directly involved. The difference is eye contact. There's no right or wrong technique here, but you need to know the impact of each; you also need to be careful not to accidentally switch from one to the other. Constant switches can be very disconcerting for an audience, as we discussed in Chapter 2.

In **FIGURE 7.23**, notice how her simply holding a prop—a wine glass—makes the image more believable. Second, see how the image of Allison looking down has a feeling of isolation and desolation, while the image where we see her eyes creates a feeling of desperation.

FIGURE 7.22 Compare the emotional difference in watching someone reflect over a drink versus looking you directly in the eye.

FIGURE 7.23 Note, first, the emotional difference when we see Allison's eyes and, second, her use of a prop.

Where possible, give your talent props to work with. It gives them something to do with their hands and provides them a place to focus their eyes and actions. And don't overlook the power of someone's eyes to tell your story.

Finally, look at the lighting. By intent, her face is dark, her body strongly rim lit. The glass sparkles; she does not. There's a story just in her sitting, desperate and alone, at an empty bar.

I call **FIGURE 7.24** the "Greek chorus" effect. Regardless of what the presenter is doing in the foreground, the audience will take its cue from the person in the background *reacting* to what the presenter is doing. The background doesn't even need to be well lit; Kim isn't in these shots. This effect is so powerful it's why you see so many reaction shots of waving flags and nodding heads when someone is doing a testimonial ad or political spot.

It is not enough for the spokesperson to make a statement. For us to believe what we are hearing, we need to see a "disinterested party" agreeing or disagreeing with the presenter. The action in the background validates, or contradicts, the presenter *far* more strongly than the presenter's words. A person doing the wrong thing in the background will destroy the credibility of even the most believable presenter.

This reminds me of something I see constantly at tourist attractions that just drives me *nuts*.

FIGURE 7.24 Which makes the product appear more trustworthy: where the person behind is unhappy or happy? (I call this the "Greek chorus" effect.)

At every tourist site in the world, people stand next to something far larger than themselves, while the person holding the camera backs up into the next county to be able to get both the object and the person standing next to it in the shot (**FIGURE 7.25**). Except that now the person in the shot is about the size of a pea and unrecognizable!

This is such an important concept that I recruited a local 6-year-old to illustrate this point. In the left image Sophie is standing next to her school sign. (Virtually every proud parent has an image similar to this posted somewhere in their house.) In the right image, though, we leave the framing of the sign

the same and move our star student closer to the camera. Makes a big difference, doesn't it? Now we see both the sign *and* a lovely close-up of her face. (If you want to get fancy, adjust the depth of field on your camera so both are in focus.)

Remember, talent have feet. There's no rule that says they need to stand *next* to something. In most cases, they look much better standing well in *front* of it.

FIGURE 7.26 is the same scene from three different angles—wide, medium, and tight. The wide shot sets the scene, the medium shot brings us into the conversation, while the close-up is all about reaction and emotion. We can use all three to tell the story.

(As a production note, Kim's face and eyes are the brightest part of the frame, while Allison's face is turned toward the camera so we can see her reaction even though we are shooting from behind her.)

Cameras today take great pictures. Exposure and color are most often correct, and autofocus lenses make sure the right part of the image is in focus. All we really need to do is aim and shoot. But, within that "aiming" lies a world of possibilities.

The more you think about lighting, blocking, composition, and framing, the better—and more persuasive—your images will look.

FIGURE 7.26 This is the same scene except for the camera angle. Note how the emotional impact changes.

KEY POINTS

Here are the key points from this chapter:

- Creativity is a collaborative enterprise between all team members.
- All images, still and moving, are covered by copyright.
- All images need to start with a plan covering the message, audience, story, and image.
- There are four creative components of a photograph: lighting, blocking, composition, and framing.
- Control of light is essential to all images.
- There are an unlimited number of options in how to light a scene.
- Three-point lighting is a way to create the illusion of depth in a two-dimensional image.
- The 180° Rule helps plan where to put cameras in relation to talent.
- Always talk with your talent. Explain what's going on, encourage them in their work, and talk to them first at the end of a shot.
- Our goal, in any persuasive work, is to capture the eye of the viewer long enough to deliver our message.

PRACTICE PERSUASION

Using your cell phone, take photos of a friend while changing the position of the lighting (room lights are fine). See how changing the angle of the light changes the impact of the photo. Then, move outside and change their position relative to the sun.

Compare the results. Which shots are more interesting? Which lighting causes you to want to look at an image again?

PERSUASION P-O-V

IMPROVED MEDICAL COLLABORATION

Lawrence P. Kerr, MD
Cheryl B. Kerr, MD
Founders of ClickCare, www.clickcare.com

Medicine is a victim of the digital age, not the medicine of the biller, the payor, or the regulator. They are the victors. The loser is the medicine for each of us—the medicine for the patient and the medicine for all who strive to care for us. The scientific knowledge base is blooming. Exciting discoveries are announced daily. We are collecting massive amounts of data while praying that artificial intelligence will save us from drowning in it.

Yet, the most common medical error is a failure to communicate and collaborate; communication lost between silos of expertise. "Store and forward" telemedicine with a video clip solves this and shows the great importance of the moving image.

Initially, medical conditions were described by often revolting analogs of food such as a "bunch of grapes" or "a pancaked-sized tumor." Understanding now is inhibited by forms with check boxes when it needs to be enhanced by rich data. There is no data that is richer than the moving image. However, live video conferencing solutions, though immediate, are highly interruptive and are difficult to use for collaboration and review. They are not stored-and-forwarded.

I'd like to present another example of the use of a video clip. A seemingly healthy 15-year-old comes into the pediatrician's office. She can't walk; one foot is dragging. Early that morning, she was at swim practice and doing fine. What is the problem? Who should be called? Is it a back problem, a nerve problem, a muscle problem? Is it in the brain, the spinal cord, the leg, or even the mind?

These areas are generally outside the expertise of the pediatrician and are obviously rare. Still, the patient and parent both want help, but who should be called? The specialties (orthopedics, neurology, neurosurgery) that might treat this problem overlap, and all have long waiting lists. The patient should not be told to go see one of them and be sent home. Better help than that is needed, and given the rapidity of the change, that help should be soon. Telephone tag and simplifying the problem to the label of "foot drop" is the standard of care but far from an adequate/efficient/elegant telling of the story.

Using a short video clip and a focused history, all three consultants can be selected. Because the video clip was stored and accessible (through a secure connection), each looked and made a comment.

Orthopedist: "I see no, nor did I understand, any history of injury. It will be a month, or so, until I can see the patient, but I don't think I will be much help. I would like to read what the others say."

Pediatric neurologist: "I'll try to be available but have had to financially curtail some of my practice. It will be seven weeks until I can see her. I can discuss this with the others, though."

Neurosurgeon: "I have instructed my secretary to ensure I see her tomorrow. I am changing my surgical schedule for next week. This problem must be taken care of right away. A biopsy is necessary, but this is most likely a benign, rapidly growing tumor in the spinal cord. I am calling for the radiologic studies now."

Indeed, the presumptive diagnosis was correct. The patient was discharged after an outpatient stay, was saved from permanent paralysis, and more than $40,000 was saved in unnecessary expenses.

Using video to tell and to share a medical story is different than other filmmaking.

The first principle is that the data (the video clip) cannot be modified. While the editing tools are identical, their use is different, almost completely reversed. In a hospital or clinic setting light needs to be adapted to, rather than set up. Mood and colorization cannot be modified. Special effects are to be used only to highlight or bring focus. Framing is standardized both for orientation and for comparison over time.

The second principle is that every effort must be made to tell the story as quickly as possible. The time constraints of the viewer are respected. The storage requirements of the system are limited. The transmission times are important constraints.

The third principle relates to the creator. Production needs to be fast and simple. Distribution similarly needs to be secure. The privacy of the patient is important legally (HIPAA in the United States) and ethically.

As creators, we want to be sure we are showing the real over the ideal. We have found that a 15-second story is all that is needed. The viewer's attention is not lost, the storage is smaller, and the transmission time is satisfactory. As creators, we need to understand the problem we are communicating. In the case presented, both right and left legs are important to contrast the difference in symmetry, and a few steps in motion are important. HIPAA constraints dictate that identifying characteristics be left out. In this case, the face is not important, but the lower body clearly needs to be shown.

We live in exciting times. We have powerful tools. We need to bring our skills to help people both inside of healthcare and in the rest of life. In a way, it is all the same.

To err is human,

To edit, divine.

SRITHIP SRESTHAPHUNLARP

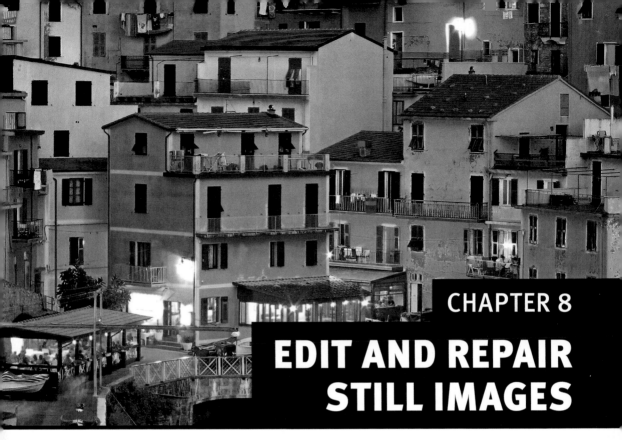

CHAPTER 8

EDIT AND REPAIR STILL IMAGES

CHAPTER GOALS

Once a photo is captured, there are still many things we can do to it to make it more persuasive. This chapter looks at how to edit and repair single-layer images, specifically, photographs. The next chapter illustrates how to create multilayer images, called "composites," which includes adding text and combining multiple images into one.

In this chapter, we will:

- Cover the ethics of editing an image
- Learn the basics of Adobe Photoshop
- Open and edit images
- Repair images
- Adjust images
- Save images

OFTEN, THE EASIEST WAY to learn Photoshop is to manipulate an existing image. For example, you may need to get rid of unappealing elements, improve the exposure, add text, or combine multiple images—whatever is necessary to attract the eye of our audience. All of these require image editing, and image editing means Photoshop, which, for many of us, is an intimidating program.

If you've been discouraged looking at Photoshop—saying "I'm not an artist"—don't give up. I'm not an artist either. In the hands of someone who can draw, Photoshop creates *amazing* images. We aren't artists. We don't need to be. What we are is *communicators*, and Photoshop has tools that we can use every day to improve the clarity and power of our images.

Years of disheartening experience have convinced me that I can't draw a straight line without mechanical assistance. Yet, I still use Photoshop every day to create or edit images for articles and webinars to teach people around the world. You can too.

Entire books are devoted to Photoshop. Two chapters in this one does not equal a book. Instead, I want to share the basic tools you can use in Photoshop to edit, enhance, and repair images. If Photoshop catches your fancy, there are many other books that can take you as far as you want to go.

The Ethics of Image Editing

Before we start pushing pixels, I need to stress that editing an image carries with it ethical connotations. First is copyright, which we discussed in Chapter 7, "Persuasive Photos." Make sure you have the right to edit an image.

Second is content. The development team at Adobe and I have frequent debates about the ethics of image and audio editing. Adobe makes incredibly powerful tools that allow us to alter images in ways that may not be appropriate, legal, or fair. Adobe's point of view is that "they just make tools." My point is that a lot of the "fake news" we are dealing with today is because these tools are so accessible. There's no easy answer, though I think Silicon Valley, in general, too often treats ethics as "someone else's problem."

Andy Grove, the former head of Intel, once said: "A fundamental rule in technology says that whatever can be done will be done.¹" However, that doesn't give us carte blanche to do whatever we want with it.

Journalism ethics require no image editing to any image of any sort. Creating images for advertising frequently requires significant editing. My work is somewhere in the middle. In the photos I shot for this book, I cleaned up blemishes and distracting elements because I wanted to make a certain point. I think that's ethical. Using the same tools, I could put one actor's face on another's body; or put someone in a location they never visited in real-life. I think these are not ethical. These choices are up to you. I simply want to wave a flag and say, "Please consider the ethics of what you are doing. Would you like someone else to mistreat images of you?"

Please consider the ethics of what you are doing.

Getting Started with Photoshop

Before opening the app, let's review the foundation upon which all digital images are built: bitmaps.

Bitmap Fundamentals

A bitmap, as shown in **FIGURE 8.1**, is a rectangular grid of generally square pixels where each pixel is the same size as its neighbor, and each pixel has one and only one color. Photoshop, at its simplest, is a pixel editor. (In contrast, Illustrator, at its simplest, edits curves called "vectors." Vectors are highly useful visual tools, but for this book, I am concentrating on bitmaps.)

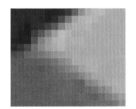

FIGURE 8.1 An enlarged detail from a photo bitmap. *All* digital images are composed of rectangular bitmaps.

Pixel

Short for "picture element," a pixel is the smallest component of any digital image, moving or still. Pixels are fixed in size and aspect ratio, contain one and only one color, and can be transparent, translucent, or opaque. Pixels are organized into a rectangular grid called a bitmap. Everything we do in Photoshop comes down to manipulating pixels. Almost all pixels in digital images are squares.

The big benefit to bitmaps is that they display quickly even on older systems, are easy to edit, and provide lots of color and texture. The limitation of bitmaps is that they create large files, sometimes *very* large files, and that they can't be enlarged without getting blurry.

Each pixel can have up to four values associated with it: the amount of red, blue, and green that make up the color, plus the amount of transparency. While a pixel has no default color, all pixels by default are fully opaque.

[1] (Attributed to Andy Grove in: Ciarán Parker (2006) *The Thinkers 50: The World's Most Influential Business.* p. 70)

When you open Photoshop, you see the interface in **FIGURE 8.2**, though, ah, without the image. Tools are on the left, tool options across the top, a series of control panels on the right (I will talk about these in the next chapter), and a big window where our images are displayed in the center.

FIGURE 8.2 The Photoshop interface showing the Bow River near Banff, Canada.

Optimize Preferences

On Macs, Preferences are in the Photoshop menu. On Windows, Preferences are in the Edit menu.

Before we start editing images, we need to change two settings in Preferences and one menu setting to optimize Photoshop for our work.

All digital images are sized in pixels. But, by default, Photoshop measures images in inches. So, change Preferences > Units & Rulers to Pixels (**FIGURE 8.3**).

Then, to improve image quality when resizing, change Preferences > General > Image Interpolation to Bicubic Sharper (**FIGURE 8.4**). This improves edge definition when resizing lower-resolution images, preserving the apparent focus of an image.

Finally, after you open an image into Photoshop, which I'll get to in a second, go to the lower-left corner, below the image, and click the small right-pointing arrow and change the setting to Document Dimensions (**FIGURE 8.5**). This

FIGURE 8.3 Set Units & Rulers to Pixels.

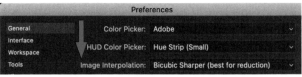

FIGURE 8.4 Set Image Interpolation to Bicubic Sharper.

displays the size of your image in pixels, with the horizontal dimension listed first. Since everything we do for the web these days is based on pixels, changing this setting helps us make sure all our images are the right size.

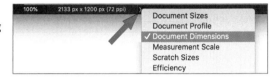

FIGURE 8.5 Switch the lower-left display to Document Dimensions.

Explore the Interface

As you might guess, to open an image, choose File > Open, the same as any other application. Then, navigate to the image you want to work on.

Other Ways to Open Files

- Drag one or more images into the interface.
- Drag one or more images onto the Photoshop dock icon.
- Double-click an image in the Finder (though this may open Preview instead of Photoshop, depending upon settings in the Finder File > Get Info dialog).

Now that you've opened an image, let's explore the interface a bit. To zoom in, press Cmd+[plus]. To zoom out, type Cmd+[minus]. To fit the image into the screen, press Cmd+0 (that's a zero). To display the image at 100 percent size, press Cmd+1. (All these shortcuts do is change your view; you are not altering the image in any way.)

Press V to select the Move tool. This is the *most* useful tool in Photoshop. When in doubt, press V to reset your tools and switch the cursor back to its default operation.

Press Cmd+[plus] a few times to zoom in. Press the Spacebar, which displays the Hand tool, and drag the image. This allows you to change your view of the zoomed-in image.

All the tools we need to change our images are stored in the Tools panel on the left. The problem is that the position of these tools changes depending upon how you configure your workspace. It is easier to search for a tool than to track it down in the Tools panel.

Before you open your first image, or any image actually, make a copy of it. That way, if anything goes wrong, you still have the original as a backup. This is especially true if you are new to Photoshop.

All these techniques work exactly the same for Windows users. Simply substitute Ctrl for Cmd and Alt for Option. (Mac users, Cmd is short for "Command.")

Click the magnifying glass (the Search icon in **FIGURE 8.6**) to display a search window. Enter the name of the tool you need and press Return. Let me illustrate. (You can also access the Search box by selecting Edit > Search or pressing Cmd+F.)

Getting Started Editing

The best way to learn Photoshop is to edit an existing image. That gives us something to work with, without worrying about how to create it in the first place. Open the image of your choice using File > Open. Then, let's start exploring by correcting errors we made during shooting.

Straighten an Image

Most of the time, when shooting stills, we hold the camera in our hands. This means that, often, our images may not be level (**FIGURE 8.7**). This needs to be fixed because, while we may *shoot* tilted pictures, our eyes don't like to *look* at tilted pictures.

Fortunately, Photoshop has a hidden tool that allows us to easily straighten an image. Click the Search icon in the upper-right corner and enter "Ruler Tool"; then press Return. The Ruler tool is highlighted in the Tools panel (Figure 8.7). See those small arrows in the lower-right corner of many tools? Click one and discover multiple tools hidden under one icon. This is why I like using the Search box—I can never remember which tool is where.

Next, find a line in your image that should be horizontal or vertical. Click and drag the Ruler tool along that line. (In this example, I raised the line slightly above the horizon so you can see it more easily. It should actually track along the horizon.) This tells Photoshop the horizon should be level, but isn't. Finally, in the Options bar, click the Straighten Layer button. Photoshop rotates the image and removes the tilt.

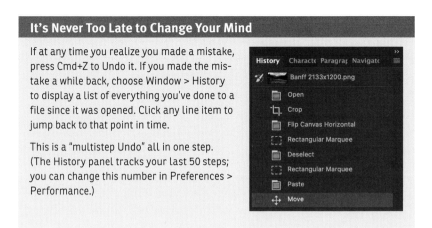
The trick to using the Ruler tool is to find something in your shot that is actually supposed to be horizontal or vertical. If the results don't look right, Undo; then select a different edge and try again.

I find myself leveling a lot of my handheld shots.

Scale an Image

The next task we often need to do is scale, or resize, an image. Most digital cameras shoot images measured in "megapixels," or millions of pixels. These are far too large to post to the web, any web page that contained them would take forever to load. So, we need to scale these to a smaller size.

Remember, scaling any bitmap larger than 100 percent will always make it look soft, blurry, and, generally, bad.

Go to Image > Image Size to display the dialog shown in **FIGURE 8.8**. This changes the size of your entire image. Notice the Resolution setting at the bottom? As mentioned in Chapter 5, "Persuasive Colors," "resolution" can

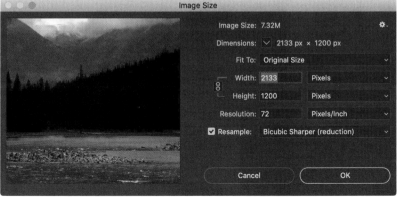

FIGURE 8.8 The Image Size dialog allows you to change the size of an image.

mean either the number of pixels in an image or pixel density. In this case, resolution means pixel density (pixels/inch), which applies *only* to images you print.

For all images posted to the web or digital video, you can ignore the Pixels/Inch setting. Instead, pay attention to the total number of pixels across (Width) and high (Height). As well, Bicubic Sharper is the best option for scaling lower-resolution images. (This is why I changed the Preferences settings earlier in this chapter.)

If you click the Pixels menu, you'll see options that allow you to enter either a pixel or a percentage value for scaling the image. While I generally use the Pixels option, it's nice to know that we can also use percent when needed. (For example, on my website, all my images need to be no wider than 680 pixels; the height doesn't matter. So I scale many of my screen captures to fit within this horizontal width limitation.)

Mac Retina Displays

This image represents the higher pixel density of screen shots taken on Mac Retina displays (called HiDPI on Windows). When posting images to the web, ignore the Resolution setting and just look at the total pixels for Width and Height. For web work, I set this to 72, just to keep all my images consistent.

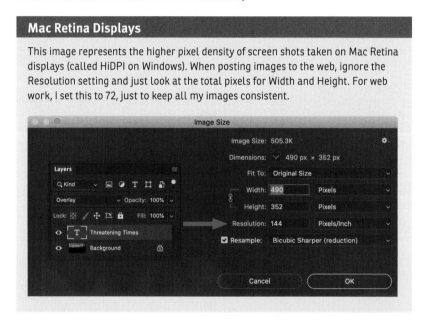

Crop an Image

The process of selecting and retaining just a portion of the image is called "cropping." A good idea, during your photo shoot, is leave a little more space around the edges than you need. This simplifies straightening images as well as cropping. It is easy, as you are about to learn, to crop images to remove what you don't need. But, if you accidentally fail to include someone's head in the original shot, there's not a lot you can do to fix it.

As Annabelle Lau, one of my teaching assistants, told me after reading this paragraph, "It's truly a loss. I have a ton of cut-off shoes and feet because you don't think about this when you take the picture."[1]

To crop an image, search for "Crop tool" (**FIGURE 8.9**). Then, drag the highlighted corners of your image to select just a portion of the frame. (And, did you notice that Photoshop automatically supplied a Rule of Thirds grid to help guide your framing?)

FIGURE 8.9 The Crop tool (left) and a crop in progress.

In Figure 8.9, I'm cropping a portion of my photo of the Bow River in Alberta, Canada. The Crop tool highlights that part of the image we will keep, while obscuring, but not yet deleting, the portion we'll remove.

More Crop Tool Options

After dragging the corners of an active crop rectangle, click inside its boundary and drag to change the position of the image within the crop window. Press Option while dragging a crop corner to have all four sides move at the same time. Press Shift to constrain the aspect ratio.

Because cropping selects only a portion of an image, the default Crop tool options *always* change the pixel dimensions of an image. Most of the images of our two actors in Chapter 7 were cropped slightly to allow me to emphasize specific points in each shot.

If you don't have strong horizontal or vertical lines in an image, the Crop tool provides another way to level an image. After you've started a crop, move the Crop tool just outside a corner of the crop window, as shown in **FIGURE 8.10**. A two-headed curved arrow appears, allowing you to rotate the image. The white numbers against a black background show you how far you are rotating the image.

[1] Personal email.

FIGURE 8.10 Click outside the cropping rectangle and drag to rotate an image. The grid lines in the Crop window provide another way to level an image, this time using your eyes.

When I write one of my online tutorials, it's not unusual for me to include 20 to 30 images. Each one gets opened into Photoshop, and almost all get scaled and cropped. I use these tools daily.

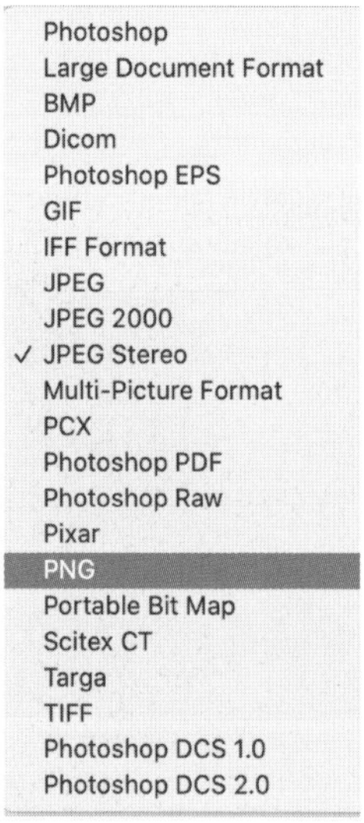

Save an Image

Just as you need to open an image before you can edit it, you need to save an image before you can send it anywhere.

If you choose File > Save, it saves your revised file using the same name and image format (just as you'd expect if you were saving a Word file).

However, if you choose File > Save As then click the Format menu illustrated in **FIGURE 8.11**, then, yup, those pesky codecs came back to confront us with more decisions.

Photoshop
Large Document Format
BMP
Dicom
Photoshop EPS
GIF
IFF Format
JPEG
JPEG 2000
✓ JPEG Stereo
Multi-Picture Format
PCX
Photoshop PDF
Photoshop Raw
Pixar
PNG
Portable Bit Map
Scitex CT
Targa
TIFF
Photoshop DCS 1.0
Photoshop DCS 2.0

FIGURE 8.11 Photoshop can save in a wide variety of image formats.

Cloud Documents

New in Photoshop is a button called "Save to cloud documents." Cloud documents are Adobe's new cloud-native document file type that can be accessed online or offline directly from within the Photoshop application. Cloud documents can be accessed across devices, while your edits are automatically saved through the cloud. Principally designed for mobile devices, this format is supported in all versions of Photoshop. Cloud documents are not supported on Windows.

> Save to cloud documents

TABLE 8.1 helps you choose the right file format for your image.

TABLE 8.1 Select the Right File Format		
TASK	**FILE FORMAT**	**NOTES**
Save a compressed file to post to the web.	JPEG	Set image quality between 6–10.
Save a high-quality master file.	TIFF	TIFF is high-quality, but support for layers and transparency is not widespread.
Save a high-quality master file, with a smaller file size than TIFF.	PNG	Transparency in PNGs is possible, but not widely supported.
Save a high-quality file for later editing with layers, transparency, and wide application support.	Photoshop (PSD)	This keeps all elements separate, allowing future changes.

My preference is to save master files that I might re-edit as PSDs, master files where all elements are saved as a single layer to TIFF or PNG, and all files for the web as JPEG. You can, of course, pick other options, but these options cover all the work that I do on a regular basis.

Repairing an Image

To me, the difference between "editing" an image and "repairing" it is based upon the condition of the image. However, as I write this, I find myself using all these techniques on both old and new images, so perhaps this definition is based on what's in my mind more than in reality.

Most of the time, repairs are simple: remove a stray piece of hair or a scratch in an old photo, adjust exposures or color balance, or remove a distracting element in the background. This section shows how these are done.

The Spot Healing Brush

FIGURE 8.12 is a typical repair example; it shows an older photo with some damage. There's a big white spot near his pocket, a tear near the right side of his neck, and dirt in the middle of his forehead. All of these can be cleaned up using the same tool: the Spot Healing Brush.

FIGURE 8.12 This Charles Faben, Jr., image needs repair to remove blemishes.

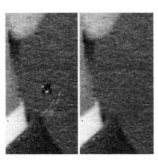

FIGURE 8.13 The Spot Healing Brush (left) and image damage before (middle) and after repair (right).

Search for "Spot," and Photoshop displays the Spot Healing Brush.

Select the Spot Healing Brush (**FIGURE 8.13**) then move the cursor on top of the blemish you want to fix. A small circle appears at the tip of your cursor. Adjust the size of the circle to be a bit larger than the blemish you are trying to fix. (Type the left or right square brackets to increase or decrease the size of the circle.) Click the blemish and—poof!—gone. Photoshop matches the color and texture of the area around the repair and works its magic.

In Figure 8.13, I used the Spot Healing Brush to remove the hole in the image and the scratch just below it. I use this constantly to repair/remove scratches, blemishes, background wires, and stray hair. It's an amazing tool!

Remove Redeye

When taking a quick photo using a camera light, all too often the subject's eyes glow red. This is caused by light from the camera bouncing off the retina in the back of the eye.

Photoshop has an easy way to fix this. Go to the search box and look for "Red Eye Tool." It's in the same menu as the Spot Healing Brush. Select the tool then drag a selection rectangle around each red eye. Poof! Gone.

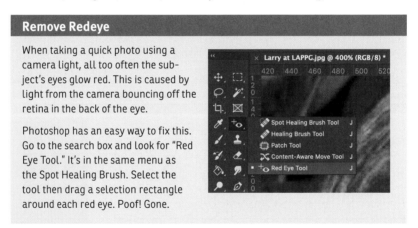

The Clone Tool

The Spot Healing Brush makes mistakes disappear. The Clone tool allows you to replicate a portion of an image somewhere else. By now, you know how to find a tool—search for its name.

To use the Clone tool, press the Option key and click somewhere identifiable. This selects what you want to copy. (In **FIGURE 8.14** I used the intersection of the tree trunk on the right with the flowers at the edge of the hillside.) Then, click where you want to paint the copied image and drag, as though you were painting. A new tree grows instantly.

FIGURE 8.14 The Clone tool copies a portion of an image somewhere else, as we are copying the tree on the right to create a small wooded area.

The trick to using the clone tool is finding the right reference so perspective and horizon lines match. By Option-clicking the edge of the flowers, it becomes easy to paint the new tree starting at the same edge. Using the Clone tool to create believable copies takes a bit of practice, but, with practice, this becomes a useful tool for removing overhead power lines from a scenic shot, adding shrubbery, or removing a wall plate from behind talent (**FIGURE 8.15**).

The Patch Tool

The Patch tool allows us to get rid of areas we don't like—for example, the white wall plates on the right side of Figure 8.15. Select the Patch tool, then drag to create a small selection circle around the area you want to remove. (A "selection" is a region defined by a dotted line. We'll use them a lot in the next chapter.) As you drag, the selected region is replaced by the region you drag the selection area into.

In Figure 8.15, I used the Clone tool to remove the wall plate by Allison's shoulder, because I needed the precision it provided. Next, I used the Patch tool to remove the two units higher on the wall because the Patch tool works better with areas than the Clone tool. Why am I removing these areas? Because they distracted the eye from what I wanted you to see, which was the dialogue between Allison and Kim. There was no reason to keep those distractions in the shot.

To summarize, the Spot Healing Brush replaces a blemish with pixels that match the surrounding area. The Clone tool allows us to copy portions of the image from one place to another. The Patch tool allows us to replace one region with another. I used these tools frequently for the images in this book.

FIGURE 8.15 These wall plates are a good example of distraction. I used the Clone and Patch tools to remove them.

Adjust Image Exposure

Like many older photos, the picture in **FIGURE 8.16** is a bit washed out. (It also needs straightening and cropping, which we already know how to do.) One of my favorite tools in Photoshop—and one of the oldest—allows us to adjust the exposure: Levels.

Exposure is another way to say "grayscale values." Grayscale is what makes an image light or dark. It also provides texture. Adjust grayscale values first, before you adjust colors.

The Levels panel (**FIGURE 8.17**) allows us to adjust grayscale values for an image or a selection, which we'll cover in the next chapter. The Histogram, which we first met in Chapter 2, "Persuasive Visuals," displays the range of grayscale values for the entire image, with black on the left, white on the right, and midtone grays in the middle.

FIGURE 8.16 Here's what the well-dressed 4-year-old wore in Toledo, Ohio, in 1884. The left image is the source; the right image is after rotation and cropping. (And, yes, I love that little lamb, too.)

Notice that the pixel values, represented by the white shape, don't go all the way to each end. They are bunched in the middle. This is typical for images with a "washed-out look." There are no rich shadows or highlights, just a bunch of gray in the middle.

To correct this, move the bottom-left slider (left arrow) to the right until it just touches where the histogram starts to curve up. This moves the value of the darkest pixels in the image closer to black, making the shadows fuller.

Then, drag the right slider (right arrow) until the histogram just starts to curve up. This moves the lightest pixels closer to white, adding sparkle and energy to the image.

Finally, move the middle slider until the image looks "better" to you. This is a subjective judgment on your part—there's no magic number that works for all images. **TABLE 8.2** summarizes these adjustments.

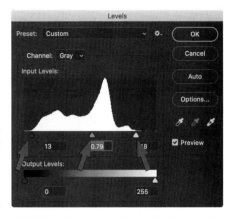

FIGURE 8.17 The Levels sliders adjust black (left), midtone (middle), and white grayscale values in an image or selection.

TABLE 8.2 Adjust Grayscale Values to Maximize Emotion

ACTION	RESULT
Adjust shadows (step 1.)	Makes an image feel "richer"
Adjust highlights (step 2.)	Gives an image energy and snap
Adjust midtones (as needed.)	Changes the emotional feel and/or time of day

The best way to see how this works is to play with an image. You don't need to move these sliders a lot. But, especially for a washed-out image, Levels can create a world of improvement.

Add Fog—or Take It Away

As an experiment, load a photo of an exterior scene into Photoshop, open the Levels dialog and drag the Black and White sliders for Output Levels, at the bottom, toward each other. Notice how the image gets "foggy" as the deep blacks and sparkling whites disappear? Not only can we make images look crisper, we can also make them look like they were shot in a deep fog. Cool!

FIGURE 8.18 The finished image.

Note the power of the midtones. As you slide these, especially on an exterior shot, notice how the time of day seems to change, which also affects your emotional response to the shot.

FIGURE 8.18 is the finished result, after straightening the image, cropping, and adjusting levels. See how she "pops" off the page? It's a much better image.

Adjust Color

There is an automated tool that solves another problem: incorrect color. I am not a fan of automated color tools; most of the time I prefer my own settings. But, this one can bail you out in a pinch, without you having to know anything about color theory or color correction.

Many times we need to deal with a color cast, which results in an image that's too green or blue or orange. **FIGURE 8.19** is a good example of this. Choosing Image > Auto Color will remove most of that color cast. Generally, this will get you close, but it won't equal the results of using more specialized color tools in Photoshop.

> **A Fast Way to Create Black-and-White Images**
>
> Thinking of color, to quickly make any image black-and-white, choose Image > Mode > Grayscale. When the "Discard Color" warning appears, click Discard. However, once you save that image, all the color information is removed—it will be black-and-white permanently.

FIGURE 8.19 The plane's image is too green (left). After applying Image > Auto Color, the green cast is removed, and the image looks "normal."

FIGURE 8.20 illustrates another use of these tools: not to repair an image but to create a look. Here, I adjusted the Levels to give the image a washed-out look. Then, I reduced the saturation (the amount of color) more than 60 percent using Image > Adjustments > Hue & Saturation and dragging the Saturation slider to the left.

These washed-out, desaturated images are often used as "before" images in a commercial comparing the results of using a product. Black-and-white is also popular for these Before & After comparisons.

FIGURE 8.20 An image with 50 percent saturation, that is, with half its color removed, using Image > Adjustments > Hue & Saturation.

Color Balance

There's one more tool I want to introduce to you: Color Balance (Image > Adjustments > Color Balance). This allows you to adjust the amount of red, green, and blue in the shadows, midtones, and highlights of an image. (Photoshop pros would point out that Curves take these adjustments to an entirely different level, but explaining how to use Curves is well outside the parameters of this chapter.)

In **FIGURE 8.21**, the source image of a cabin in the woods was green, not surprising with all the foliage around it. I applied the Color Balance tool, which allows us to adjust the amount of color in the shadows, midtones, and highlights. Virtually all color adjustments, especially the easy ones, are made in the midtones.

FIGURE 8.21 The source image was green. Using Color Balance, I removed some of the green from the midtones and added red.

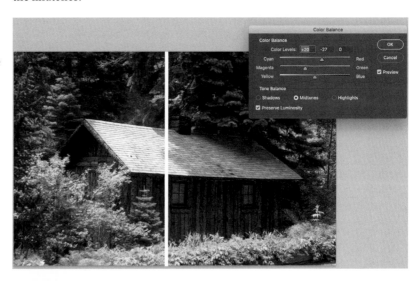

To remove a color, slide the control toward the opposite color. So, to remove green, slide the control toward magenta, which is the opposite of green. ("Opposite" refers to the color wheel presented in Chapter 5.) To increase the amount of red, drag the top slider toward red. Keep adjusting until the image looks "right" to you.

As you can imagine, there is a lot you can do with color, as you'll discover as you experiment on your own. These tools will get you started. The good news is that you don't need to be a color expert to improve your images. In most cases, we only need to make minor adjustments to an image. These tools are here to help.

KEY POINTS

In this chapter, we looked at how Photoshop can help edit and repair single-layer images. Here are the key points I want you to remember:

- Any image editing involves ethical decisions. Treat your images and the people in them with respect. .
- All digital images are bitmaps, which are a rectangular grid of pixels.
- When editing an image, the two most common improvements are straightening and cropping the image.
- When repairing or adjusting an image, the two most common fixes are the Spot Healing Brush and Levels.

PRACTICE PERSUASION

Using the image editing tools on your cell phone or computer, take a photo you shot for the exercise at the end of Chapter 7 and:

- Remove the color.
- Boost the amount of color using saturation.
- Adjust levels to make it more or less "contrasty."
- Modify the colors to make it more gold, blue or green.

How do these changes affect your interpretation of, or emotional response to, the image?

PERSUASION P-O-V

FINAL MEMORIES

Natasha Mary (and Michelle Benton)
Owners: Tasteful Transitions
www.tastefultransitions.com.au

You get only one chance to evoke a response in people using the images they provided to create an emotional funeral slideshow.

When you attend a funeral, the funeral director or family present a photo slideshow to attempt to relay the meaning and magnitude of the deceased life. Often the photos are randomly sequenced to a song with no apparent connections. This means people are not drawn into the visual story, not drawn into the emotional experience of this life. They become just photos to a song.

Family photos are the raw materials I use to create a wonderful living memorial to evoke feelings and enhance the memories of lives shared. Working with families to depict who their loved one was means conveying who they were through images and the deliberate sequencing of those images. As we go through the photos, I listen to the information and stories the family tell me in relation to who their loved one is to them. As we are talking, I move the story into a visual format using the photos we are sorting. The combination and arrangement of the photos and the experiences they are sharing provide the "magic" people witness at the end.

The family and contextual needs are different for different age groups, and the audiences have different responses and associations with images. Here are four examples.

The death of a baby is always difficult. In this world of cell phones, even a short life is well-documented. Olivia was a cot death baby of 3 months. She was her mother's dreamcatcher baby. Mum had a tattoo of a dreamcatcher to represent Olivia. I used photos of this and the photos of the dreamcatchers hanging in her room to weave through the photos of her to help create the story of this short but impacting life. The slideshow had some snippets of her ultrasound with sounds of her heartbeat at the beginning. I photographed her toys and things of meaning to her family. I did some of these with Mum and Dad holding each side of a "green screen" in front of her toys and memorabilia where I later inserted some photos of Olivia to give a greater depth of context to her slideshow.

Tracey was a young artist who did amazing and intricate line drawings, especially of her first love, animals. I photographed some of these artworks and wove them through parts of the slideshow. She drew a fine-lined tiger, and I faded this through to a photo of her with a tiger. There was a photo of her feeding parrots all around her, which I was able to incorporate into her amazing drawing of parrots with hearts as feathers on the wings. Her slideshow represented a major part of her eulogy for the service. This pictorial spoke to a great extent for the family.

Everyone who knew Audrey knew how much she loved lavender. She loved to look, smell, and grow it. Her garden was filled with it. Photos of Audrey's lavender featured strongly in her slideshow along with the photos of her. I was aware that bunches of lavender were going to be handed to all who attended the funeral as they entered the chapel. Consequently, the fragrance of lavender was rich throughout the service and created an especially deep sense of connection as people literally smelled her garden as they watched her slideshow.

When one of the partners in a marriage that lasted 50–60 years dies, it is important to depict the couple together and their shared experiences with an emphasis on the deceased. People in this age bracket often have those wonderful old photos in sepia, of them as a baby, old color-touched wedding photos, and some great black-and-whites. I take the opportunity to rephotograph these and heal any of the age marks and blemishes to give them the best representation in their slideshow. Many photos of this era have the people and/or the date written on the back, and I photograph that for possible inclusion with the photo. I also photograph mementos in their homes to add to the slideshow so the family can hang on to some of the treasured memories of the family home when it is lost to progress. At the end, I revisit the best ten photos of the couple to illustrate their love and commitment to each other through the decades.

I know I have done my job well creating a representation of a life through a funeral slideshow when I see the audience laugh and cry then applaud at the end.

With digital, you can change every pixel into anything you want, and you can perform any operation with the film footage that you can imagine. My brother [John Knoll] said, "This is the future of special effects in movies," and decided to teach himself computer graphics in his spare time.[1]

TOM KNOLL
CO-DEVELOPER, WITH HIS BROTHER JOHN, OF PHOTOSHOP

CHAPTER 9

CREATE COMPOSITE IMAGES

CHAPTER GOALS

In the previous chapter, we gained an introduction to Adobe Photoshop, with an emphasis on single-layer images. However, while its basic editing capability is impressive, the true power of Photoshop lies in two features we haven't covered yet: selections and layers. These features allow the software to create multilayer images, called "composites," which is what this chapter is about.

Time to learn more sophisticated techniques of image editing. Specifically:

- Add and format text
- Use layers and selections
- Place multiple images into an image to create a composite
- Use filters and effects
- Use blend modes

COMPOSITING creates a new image by combining two or more images into one.

"COMPOSITING" MEANS to create a new image by combining two or more images into one. Composite images are the true magic of Photoshop. Whether the composite is as simple as adding text to an image or as complex as creating a fantasy dragon, Photoshop can unleash your inner artist—or overwhelm you with all its possibilities. Let's start with something simple and see where it takes us.

FIGURE 9.1 These are the settings to create a new image for HD video.

Creating a New Photoshop Document

In Chapter 8, "Edit and Repair Still Images," we spent our time working with existing images because it is the easiest way to learn the software. Now, we'll discover, for example, how to create a new image to be used on the web. Starting new is useful when you need to create an image at a specific size, rather than try to scale an existing image to fit.

Choose File > New to display key settings in the New Image dialog (**FIGURE 9.1**). Give your new file a name at the top then specify its dimensions in pixels. Remember to keep Resolution at 72, simply because Pixels/Inch applies only to images that are printed.

Always use RGB for any digital image; other options are only for specialized uses. This is also where you set the color for the background. The default is white, but you can also pick black, transparent, or a custom color. I'll explain more about Photoshop's colors later in this chapter.

Another important setting is Pixel Aspect Ratio. Use Square Pixels for anything going to the web and virtually all digital video. (Standard definition, HDV, and early versions of Panasonic P2 media use nonsquare pixels.)

Adding and Formatting Text

The easiest way to illustrate compositing is to add text to an image. **FIGURE 9.2** is a composited image consisting of a background image and foreground text. (The image is near dawn on Mulholland Drive near Los Angeles early last winter. The text is Calligraphic 421 BT from Bitstream.)

Newer versions of Photoshop display pseudo-Latin placeholder text when you click the Text tool on an image. This is to show you what your current text settings look like. Just start typing and it will disappear.

FIGURE 9.2 Here's the finished result: Mulholland Drive near dawn.

Keep Text Away from the Edges

It is a good design technique to keep text away from the edges of an image; this is especially true for video or print. Since it is possible for images to be cropped during distribution, keeping a safety margin between your text and the edges of the image means that essential information, such as titles, URLs, or phone numbers, won't get lost. In Chapter 15, "Motion Graphics: Make Things Move," you'll learn about Safe Zones, which provide guides to positioning text and other essential visual elements.

Let's start by opening an image in Photoshop (File > Open). Search for and select the Text tool; Adobe calls this the "Horizontal Type tool." Its icon looks like the letter "T" (**FIGURE 9.3**). Click where you want the text to appear on your image and start typing.

FIGURE 9.3 The Horizontal Type tool has an icon that looks like the letter "T."

As you type, look over to the Layers panel on the right side of the interface. A new layer appears, above the background, that displays the first portion of your text. The Layers panel is a major feature in Photoshop because it is how we can add, rearrange, modify, and delete individual elements within the scope of the whole picture (**FIGURE 9.4**). There is no practical limit to the number of layers you can add to a single Photoshop document.

FIGURE 9.4 The Layers panel in Photoshop.

The top layer represents the foreground; the bottom layer is the background. You can change which image is in front by dragging a layer to change the "stacking order," the order of layers from top to bottom. Click the "eye" icon on the left to make a layer invisible or visible. Adjust the opacity to make a layer translucent. Select a layer and press Delete to remove it.

FIGURE 9.5 The Character panel modifies text formatting. These are the text settings I used in Figure 9.2.

With the text layer selected, look above the Layers panel and select the Character panel (**FIGURE 9.5**). If this is not displayed, open it by going to the menu bar at the top of your monitor and choosing Window > Character. The four options at the top of the panel are the most important:

- Change the font.
- Change the weight.
- Change the size.
- Change the vertical line spacing, which Adobe calls "leading."

I used a pale yellow color to reinforce the "early morning" theme. (While I'll show you how to change colors in a few pages, the color chip in Figure 9.5 is a clue.) Because this is a title, I tightened the vertical line spacing to bring the two lines closer together.

Next, let's add our first effect: a drop shadow. Again, make sure the text layer is selected and choose Layer > Layer Style > Drop Shadow.

FIGURE 9.6 illustrates my drop shadow settings for text. The key to using a drop shadow successfully is to make your text more readable, not call attention to the shadow itself. Keep the shadow opacity dark so that the edges of the text are clearly defined.

FIGURE 9.6 My preferred drop shadow settings. Opacity: 90–95%, Angle: 135°, Distance: 8–12, Spread: 5-8, and Size: 4–8.

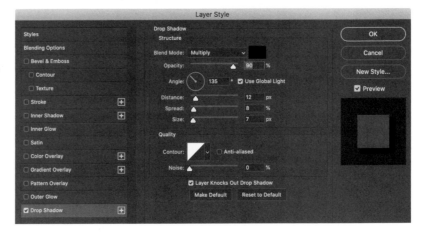

One last tweak. I didn't like the spacing between the "D" and "a" in "Daybreak," (Figure 9.2) so I kerned them by selecting the Text tool, placing the cursor *between* the two letters I wanted to adjust, then pressed Option+left/right arrow. This adjusts the spacing between just these two letters.

For Windows users, substitute Ctrl for Cmd, Alt for Option, and Backspace for Delete in these keyboard shortcuts.

Personally, I'm not a fan of effects just for the sake of effects. But as long as we're here, let's give this text a curve.

Select the text layer; then choose Type > Warp Text. This displays the dialog shown in **FIGURE 9.7**. Select a shape from the Style popup menu at the top then adjust the settings and watch what happens. Some of these are just gosh-awful, but others can be useful, depending upon your story.

LEADING adjusts the vertical line spacing between lines of text.

FIGURE 9.7 To curve text, choose Type > Warp Text and, ah, play.

Before we go further, I want to reinforce two points:

- Layers determine the order of elements from foreground to background.
- All text can be modified.

Layers vs. Background

An important point that I intentionally omitted in Chapter 8 is that when you open an image, it is placed on a special layer called the Background layer. This limits what we can do with the image.

An image placed on the Background layer can't be repositioned, nor can it easily be made transparent. For simple image editing, backgrounds are fine. But we can quickly unlock a background image and convert it into a regular layer by clicking the small lock icon (**FIGURE 9.8**, red arrow).

FIGURE 9.8 The Background layer (left) and a normal layer. Notice none of the layer controls are enabled for a background image.

Quickly unlock a background image and convert it into a regular layer by clicking the small lock icon.

For example, open a JPEG image in Photoshop. Select the Move tool (Shortcut: V) then try to drag the background image. You can't. Convert Background to a normal layer and try again. Now you can move it.

Scaling Explained

Scaling is the process of making something larger or smaller. Bitmaps, as we learned earlier, are composed of pixels, each of a fixed size and arrayed on a fixed rectangular grid. When we resize a bitmapped image, the pixels are not changing size. Instead, we are removing pixels from the image to make it smaller, or we are adding pixels to make it bigger.

This means that if we scale a layer smaller then scale it larger, first Photoshop throws away pixels to make the image smaller. Then, because those pixels are truly gone, Photoshop does its best to imagine what those pixels *might* have looked like and puts these "guessed" pixels back into the image to make it larger. Again, we are not changing the size of the pixels; we are changing the number of pixels the image contains.

FIGURE 9.9 illustrates this. The image on the left is the source, while the image on the right has been scaled to 5 percent then back. The image is blurry, details in the feathers are lost, and the sparkle is gone.

FIGURE 9.9 The Source image (left), a Smart Object scaled to 5 percent and back (middle), a bitmapped image scaled to 5 percent and back (right). Repeated scaling of bitmaps makes them blurry.

To fix this, Adobe invented a new "form" of bitmap: a Smart Object. When you scale a Smart Object, instead of using the current bitmap displayed in Photoshop (which could be missing a lot of pixels), Photoshop accesses the source file on your hard disk. Since this is always at the highest quality and not changed when Photoshop manipulates an image, this means that scaling and re-scaling will not damage image quality. The Smart Object in Figure 9.10 (middle) was scaled smaller and larger six times yet still looks like the source.

While Smart Objects are a good thing, there's a trade-off. You can't crop or apply most effects to a Smart Object. Once you have your image scaled and in place, select the layer containing the Smart Object (it has a small white square in the lower-right corner of its icon) and choose Layer > Rasterize > Smart Objects. That converts it back into an ordinary bitmap.

CAUTION: Because images in Photoshop are treated as bitmaps, enlarging any bitmap more than 100 percent of its original size will look blurry. This is true for both normal bitmaps and Smart Objects. Scaling bitmaps smaller does not damage image quality, though it may reduce detail.

Manipulating Images Using Free Transform

The Transform controls change scale, rotation, distortion, perspective, and warp. Most of the time, we'll use them to change scale or rotation. For example, after converting an image to a normal layer, choose Edit > Free Transform. (This option is not available for images opened into the Background layer.) Drag a corner to change the size of the image (**FIGURE 9.10**). Click just outside a corner and drag to rotate an image. (This is exactly the same technique we used in the previous chapter to adjust the cropping area for an image.) Click and drag the center of the image to reposition it. When you are happy with the new location, press Return. To adjust the image again, reselect Free Transform.

Free Transform can be used on any element on any layer, except the Background layer. Images can be moved, resized, and rotated as needed.

FIGURE 9.10 Modifying the size, rotation, and position of an image on a layer using Free Transform.

Place vs. Open

A fast way to place an image is to drag it from the Finder onto an open image in Photoshop.

So far, when we wanted to work with an existing image, we opened it. But, now that we know about layers, a new option is available: Place. Choosing File > Place Embedded places this new image into an empty layer above an existing image. In other words, Place creates an instant composite. A fast way to place an image is to drag it from the Finder into the existing image.

In **FIGURE 9.11**, I used Free Transform to scale and reposition the bottom image. I then added text, darkening the color of the text in the top right to make it more visible against the white background. Then, I chose File > Place Embedded and selected the Berries file. Photoshop opened and placed it into a higher layer above the background image, with a Free Transform rectangle around it. All I needed to do was scale and reposition it.

FIGURE 9.11 The bottom image was scaled using Free Transform. The top image was placed then scaled and repositioned by dragging.

Notice how the berries image overlaps the mountain image? Since the stacking order of elements in the Layers panel determines foreground versus background, the layer containing the berries image needs to go above the layer containing the mountain image. There is no limit to the number of images you can place, stack, or overlap in a Photoshop document.

Embedded vs. Linked

There are two Place options: Embedded and Linked. Unless you are working with gigantic files, Embedded is a better choice because it includes ("embeds") the image inside the Photoshop document file. We'll discuss linking when we cover video in Section 3.

Selections Create Magic

Selections are the other major feature in Photoshop. In fact, the primary interface rule for Photoshop is "select something then do something to it."

There are close to 20 different ways to select something in Photoshop. We've already glimpsed one: Crop. But cropping always changes the size of the image. What if we want the image to stay the same size but remove portions from it? That's where selections come in. For instance, we use selections to:

- Remove a portion of an image.
- Add an effect to a portion of an image.
- Move a portion of an image.

Some people call selections "masks," because they mask, or hide, a portion of an image. But, in Photoshop, selections do far more than masks. As you learn more, you'll use both. Common selection tools include:

- Marquees
- Lassos
- Select > Subject
- Object Selection, Quick Selection, and Magic Wand tools
- Quick Mask and other masking tools

Let me illustrate some of the more popular selection tools.

Near the top of the Tool panel are the geometric Marquee tools, shown in **FIGURE 9.12**. When a tool has a small (*really* small) triangle in the lower-right corner, click and hold the mouse button on the tool to reveal the other tools grouped with it. The Rectangular Marquee tool selects rectangular or square areas, while the Elliptical Marquee tool selects oval or circular areas.

Select a layer then select the Marquee tool. Drag across an element in the image. You just created a selection! Here's what we can do with it:

- Press the Delete key to remove the selected area. (Press Cmd+Z to undo your deletion.)
- Choose Select > Inverse then press the Delete key. You just deleted everything *except* the area you selected. (Press Cmd+Z to undo.)
- Try the same things using the Elliptical tool and see that it works the same way, except that this tool selects ovals or circles.

The primary interface rule for Photoshop is "Select something then do something to it."

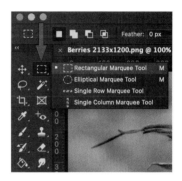

FIGURE 9.12 There are four geometric Marquee selection tools. The two most useful draw rectangles or ovals.

When using either the Rectangular Marquee tool or the Elliptical Marquee tool, press Shift to constrain the shape to a perfect square or circle. Press Option to draw from the center, rather than a corner. Press both Shift and Option to draw perfect shapes from the center.

Tweaking a Selection

If you have a selection but need to make adjustments, choose a selection tool then do the following:

- To *add* the new area to an existing selection press Shift and drag.

- To *remove* the new area from an existing selection press Option and drag.

Not all selection tools require dragging, but Shift or Option will always add to or remove from a selection.

Now that we know how to select an area, let's put this knowledge to work.

FIGURE 9.13 illustrates using the Rectangular Marquee tool to draw a selection rectangle to isolate the park bench. (Because these lines "vibrate," they are often called "marching ants.") With that area selected, I can now do something with it. For example, choose Image > Crop and only the selected area is retained. This is another way to crop an image that doesn't use the Crop tool.

Selection tools also work with the Background layer, but if you try deleting a selection from the background, you'll get an error message.

In this example, I used a selection to crop an image to reduce the size of an image. In the next example, I'll use a selection to remove a portion of an image without changing the size of the image.

FIGURE 9.14 illustrates three more selection tools: the Lasso tools.

- **Lasso tool.** This draws selections freehand (which, um, works better if you can actually draw).

- **Polygonal Lasso tool.** This creates selections by connecting a series of straight lines. To use this tool, click to set the end of each line.

- **Magnetic Lasso tool.** The selection is attracted to edges that are close to where you are drawing.

All of these tools are useful for selecting objects.

FIGURE 9.13 This image has a selection rectangle around the park bench. These lines are often called "marching ants."

FIGURE 9.14 The Lasso tools select elements that are not geometric in shape.

FIGURE 9.15 In the left image, I used the polygonal lasso to trace the rooftops of the fishing shacks. In the right image, I removed the sky, which I can then replace with a better version from a different shot.

FIGURE 9.15 shows a benefit of using a Lasso tool—we can pretty easily select odd-shaped objects. However, selecting those roof lines was tricky because of the shape of the chimney and the overlap of the roof to the walls made selection difficult.

FIGURE 9.16 highlights three more selection tools. Drag the Object Selection tool loosely around an object to select it. The Quick Selection tool selects areas based on edges. The Magic Wand tool, which is one of the oldest selection tools in Photoshop, selects similar colors.

In fact, in Figure 9.15, I cheated. While I *could* use the Polygonal Lasso tool, instead I chose the Quick Selection tool and dragged it through the sky to select the boundaries between sky and roof.

It took me a couple of minutes to create the selection using the Lasso tool. But, it took less than a second to do the same thing with the Quick Selection tool; and the Quick Selection tool did a better job! This is one of the reasons Photoshop has so many selection tools—we often need different ways to select an object.

So far, we've used selections to crop an image and remove a portion of an image. In **FIGURE 9.17**, I'm using a selection to copy a portion of an image, a balloon, to bring into another image, a mountain scene.

1. Select this balloon by clicking the Magic Wand tool on the blue sky. This selects everything blue, depending upon the Tolerance setting (see the "Tolerance" sidebar).
2. Choose Select > Invert to select everything *except* the blue sky.
3. Copy the image to the clipboard (Edit > Copy).
4. Open the image to which you want to add the balloon.
5. Choose Edit > Paste to paste the image and create a composite.
6. Drag the image to reposition it, or choose Edit > Free Transform to scale, rotate, and reposition it.

FIGURE 9.16 These tools allow you to select objects, edges, or similar colors.

FIGURE 9.17 This selection was created by clicking the Magic Wand tool on the blue sky, which selected similar blue colors.

FIGURE 9.18 illustrates a big reason to use selections: We can use them to select a portion of one image then copy/paste it into a different image to create an entirely new image. If we do it right, the images look like they belong together.

FIGURE 9.18 A composite with the balloon added to a photo of the Bow River.

When compositing, pay attention to the lighting between the composited elements. In Figure 9.18, the light direction and intensity of both the balloon and the mountain match. This makes the composite look more believable.

FIGURE 9.19 illustrates another technique: using selections to edit images. Moving elements is hard to illustrate in a book. So, for this example, I selected an edge and moved it 557 pixels to the right. It looks awful but clearly shows what moved.

Using the Rectangular Marquee tool, follow these steps:

You can't move portions of the Background layer. It needs to be converted to a normal layer first.

1. Select the area you want to move.

2. Switch to the Move tool (Shortcut: V).

3. Use Shift+left/right arrows to move the selection. (Moving a portion of an image reveals transparent pixels under each layer, which is indicated by the dark gray grid where the image used to be.)

4. To remove the transparent (blank) space on the edges, choose Image > Trim, and make sure Transparent pixels are selected.

Tolerance

Some tools, like the Magic Wand tool, have a Tolerance setting in the Options bar at the top of the interface. Tolerance determines the amount of variation from the selected color that the tool will accept. For example, setting Tolerance to 0 means *only* the selected color will be chosen. A Tolerance setting of 255 means *all* colors will be chosen. Generally, you want some Tolerance, but not a lot. If you aren't selecting all the colors you need, increase Tolerance. If you are selecting too many, decrease it. The default setting for the Magic Wand tool is 32.

Many of the screen shots posted to my website have been narrowed this way simply because they don't otherwise fit. Done right, you'll never notice the change.

Selection tools "select" a portion of an image so we can do something to it. So far, we've mostly looked at getting rid of stuff. It's time to expand into effects.

Transparent Pixels

Remember the definition of pixels? They are fixed in size, contain exactly one color, and can be transparent, translucent, or opaque. A transparent pixel is exactly that: a single pixel that we can see through.

Another powerful visual effect, as shown in **FIGURE 9.20**, is for everything to be black-and-white *except* the area you want the viewer to see first. (We also used this effect for the cover of this book.) Selections make this easy to do.

FIGURE 9.20 First, I selected the berries, then inverted the selection, then removed all the color from the selected area.

Open an image, and choose Select > Subject. Photoshop will select the dominant subject in the frame. Choose Select > Invert; this selects everything *except* the berries, in my case. Then choose Image > Adjustments > Hue & Saturation and slide the Saturation slider all the way to the left.

What the selection does is isolate an effect—in this case desaturation—to only the area bounded by the selection marquee. Here's another example.

Remember that the eye is attracted to that which is in focus. So, to make sure the background didn't distract from the title, I blurred, darkened, and desaturated it (**FIGURE 9.21**). The eye now goes first to the text, then the tree, then is free to wander hither and yon. (I'll explain how to blur a selection or layer in the next section.)

FIGURE 9.21 The background is blurred so that the eye goes first to the text, because it is brighter, then to the image of the tree.

As you can see, there are a ton of things we can do with selections. Take some time to explore the Selections menu and discover all the different options for selecting stuff. **TABLE 9.1** shows selection shortcuts I use all the time. The Appendix has more.

TABLE 9.1 Shortcut Summary	
SHORTCUT	**WHAT IT DOES**
Cmd+A	Selects entire layer
Cmd+D	Deselects selection
Shift+Cmd+I	Selects the inverse of the selection
V	Chooses the Move tool
M	Chooses the last-used Marquee tool
L	Chooses the last-used Lasso tool

Filters and Effects

The cool thing about filters is that if the effect looks good to you, it is good. There's no technical spec you need to meet. Each image and story is different, which is why there are so many different effects.

At the top of the Filter menu is the Filter Gallery (**FIGURE 9.22**). This contains hundreds of stylistic effects to change the look of an image. The easiest way to explore filter effects is to open an image, select something, apply a filter, then tweak with wild abandon. I don't use these a lot because most of my work is to repair and highlight sections of an image, rather than create something artistic.

FIGURE 9.22 Choosing Filter > Filter Gallery provides hundreds of different looks. This is the Glass effect.

The lower half of the Filter menu provides other effects options. The one I find most useful—as I illustrated in Figure 9.21—blurs a background to make the foreground more visible: Gaussian Blur (**FIGURE 9.23**). I'm always looking for ways to direct where the eye looks, and a blur creates the illusion of depth of field. The amount of the blur is determined by the Radius setting.

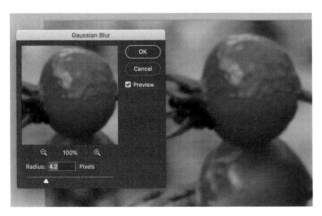

FIGURE 9.23 The Gaussian Blur dialog (choose Filter > Blur > Gaussian Blur). Blur settings are previewed immediately in the main image but not applied until you click the OK button.

FIGURE 9.24 The left example shows text superimposed over an in-focus image. In the right image, I blurred the bottom half of the image, which makes the text more prominent.

Why Is There a Small Preview Window?

The small preview window in many filters is a hold-over from the days of slower computers. Photoshop could render a small preview quickly, allowing you to see what the effect will look like once you approve it. I still remember the days of waiting 30 seconds for Photoshop to blur even a small image. If you don't see a preview in the larger window, be sure Preview is checked.

What Is Feathering?

Feathering means to soften an edge. It's an old photo-retouching term. The colors on the left have a hard edge, Feather = 0. The colors on the right have a feathered edge, Feather = 200. The amount of feathering can be varied from none to a lot.

See how, in **FIGURE 9.24**, blurring the background of the image makes the text more readable? Blurring means that nothing in the background competes with the text. This technique of blurring backgrounds, or portions of backgrounds, is one that I use all the time.

To achieve this effect, after opening an image and adding the text:

1. Make sure the lower layer is selected. In this case, it's the background layer.

2. Use the rectangular marquee to select the bottom half of the screen.

3. Choose Select > Modify > Feather to soften the edges of the selection.

4. The larger the feather value, the softer the edge. This example uses a Feather value of 150.

Now, when your eye looks at the image, it sees the in-focus area at the top and the text at the bottom, without a busy background distracting you from reading the text. Even better, your eye never notices that the background is blurred behind the text because the transition is feathered.

Choosing a Color

There are two main color pickers in Photoshop: Foreground color and Background color. (As well, text and other effects have their own color pickers. They all operate the same way.)

Photoshop provides a variety of ways to select colors, but they all start by clicking a color chip. The color chips are located in the middle of the Tools panel, as shown in **FIGURE 9.25**. Within the dialog itself, click and drag within the big color block to choose hue and saturation. Drag the small white triangles next to the green bar, in my example, to modify brightness. (You can also enter a specific color value using the numbers on the right. Professional designers use these all the time to precisely match colors.)

You can "nudge" a color by selecting a numeric value then pressing the up/down arrow keys.

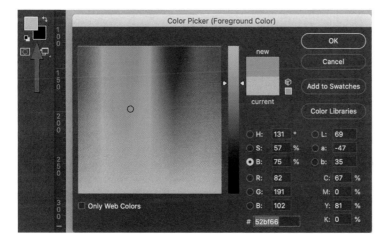

FIGURE 9.25 Color chips are located in the Tools panel. The color chip on the left picks the foreground color; the chip on the right picks the background color. To display the Color Picker, click once on either chip.

A faster way to pick a color is to open the Color Picker then click your cursor on any color in an image in Photoshop. That color is instantly "picked." Click OK to close this dialog.

TABLE 9.2 shows some handy shortcuts to working with the Color Pickers.

TABLE 9.2 Color Picker Shortcuts

SHORTCUT	WHAT IT DOES
D	Resets both Color Pickers to their default colors (black-and-white)
X	Swaps positions between foreground and background color
Cmd+A	Selects all of a layer
Option+Delete	Fills the selected area with the foreground color
Cmd+Delete	Fills the selected area with the background color

Adding a Background Behind a Layer

Let's say you opened an image full-screen, which put it on the Background layer, then decided you wanted to scale it smaller and put it against a white background (**FIGURE 9.26**). How do you create a white background? Easy. Here's how:

A good color resource is: color.adobe.com.

1. Convert the Background layer to a normal layer so you can put a new layer below it.

2. Choose Layer > New Fill Layer > Solid Color.

3. Name the layer.

4. Click OK.

5. Select the color you want from the Color Picker.

6. Click OK.

7. New layers are always created on top. Drag the layer below the current image.

This seems like a lot of steps, but after you've done it a couple of times, it will seem second nature. We are always adding new layers, always changing their order, and always selecting stuff. The only new thing we did here is to create a layer of color.

FIGURE 9.26 Create a new layer.

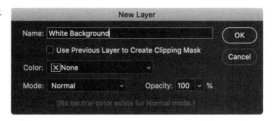

Blend Modes

Blend modes are a special kind of effect that combine textures between layers. They are often used with text, especially with motion graphics, which I'll cover in Section 3, "Persuasive Moving Images." I often use blend modes when I want to blend white text with a background.

Blend modes are applied separately to each layer, as shown in **FIGURE 9.27**. You can apply only one blend mode per layer, but every layer can have a different blend mode applied. Select the layer then choose the option you want from the blend mode menu (indicated with a red arrow in Figure 9.27).

FIGURE 9.27 Blend modes are applied to all the elements on a layer, using this popup menu. Here, I'm applying the Overlay blend mode.

There are a wide variety of blend modes to choose from, grouped into categories. Blend modes apply simple math to combine gray-scale or color pixel values between images on adjoining layers. Unlike a filter, there are no settings to adjust. **FIGURE 9.28** illustrates the blend mode options in Photoshop:

Multiply category. Combines elements based upon shadow pixel values. My favorite choice within this category is Multiply.

Screen category. Combines elements based upon the highlight pixel values. My favorite choice in this category is Screen.

Overlay category. Combines elements based upon midtone pixel values. My favorite choices in this category are Overlay and Soft Light.

Difference category. Combines elements based upon color pixel values. Each of these creates a very surreal look. My favorite choice is Difference.

Hue category. Combines elements based, mostly, upon color values, but in a different fashion from the Difference category.

In **FIGURE 9.29**, I'm using darker text placed above a dramatic sky, then applying four different blend modes to create four different effects. You can also vary results by changing the brightness, color, or opacity of the source text or background image.

In all cases, Blend modes combine the textures of the foreground and background. This makes text look more organic and part of the environment, rather than pasted in front of an image. Screen combines brighter pixels, Multiply combines darker pixels, and Overlay combines pixels in the middle. Difference displays the color 180° opposite to the source color on a color wheel (see Chapter 5, "Persuasive Colors").

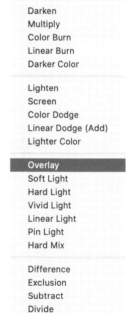

FIGURE 9.28 Blend modes are grouped into categories, based upon how they manipulate pixel values.

FIGURE 9.29 Source image and text (top), followed by four blend modes: Screen, Overlay, Multiply, and Difference (from top to bottom). With all blend modes, the texture of the lower image is combined with the text.

Blend modes are a special kind of effect that combine textures between layers.

However, blend modes can do much more. Here are some additional examples.

Blend modes combine textures between layers so that they look more organic, more "integrated." In **FIGURE 9.30**, for instance, the text looks like it was spray-painted on the background, rather than floating above it. This uses the Overlay blend mode.

We can also use blend modes to replace the solid color of the text with something more visually intriguing, including video or animation (**FIGURE 9.31**). The specific technique to create this blend mode varies by application.

FIGURE 9.30 The text on the left, being a solid color, is pretty stark. Look how much more interesting it is when it picks up the texture from the background.

FIGURE 9.31 Another blend mode replaces the solid white color of the text with a more interesting color and texture.

FIGURE 9.32 By stacking a gradient on a layer above this image then applying the Overlay blend mode, we can improve most outdoor images (before the effect on the left, after on the right). (Image courtesy of EditStock.com)

In **FIGURE 9.32**, we stacked a black-to-white gradient on top of a photo then applied the Overlay blend mode to the top image. (The gradient was created in Photoshop using the Gradient tool.) The darker values at the top darkened the sky, making the clouds more dramatic, while lightening the foreground and making it more "present." Gradients and blend modes can enhance almost every landscape image.

What makes blend modes so much fun to work with is that there are no settings to adjust. They are simple mathematical calculations performed on each pixel. Either you like the results or you don't. Instead of tweaking a filter, try different blend mode settings, modify the opacity, or slightly adjust the levels for the image.

I use blend modes all the time to make two separate images look related, rather than merely superimposed on each other.

What makes blend modes so much fun to work with is that there are no settings to adjust.

One Last Effect

As we wrap up this chapter on Photoshop, I want to illustrate an effect that I assign to my students: creating a cast shadow. **FIGURE 9.33** shows the finished results. This highlights a variety of techniques we've learned over the past two chapters.

FIGURE 9.33 Here's the finished effect, a glass that casts a soft shadow; though that shadow was not in the original image.

FIGURE 9.34 Here's where we are now—the image on the bottom (it doesn't need to be converted to a layer), an empty layer (which will hold the shadow), and the copied subject image at the top.

Here are the steps:

1. Open an image with a strong foreground object and bland background.
2. Choose Select > Subject. This puts a marquee around the foreground object. (This doesn't need to be perfect, but some tweaking may be necessary.)
3. Copy the selection to the clipboard (Edit > Copy).
4. Paste the selection. This automatically creates a new layer, places the pasted object on it, and matches its original position.
5. Create a new layer and drag it between the image and the subject in the Layers panel (**FIGURE 9.34**).

The stacking order of the layers is important, as you'll see. (I renamed the layers in Figure 9.34 to make it easier to see what's going on. To rename a layer, double-click the layer name.)

1. Rename the middle layer to "Shadow."
2. Press D to reset the Foreground and Background color pickers to black and white, respectively.
3. Cmd+click the icon in the Subject layer to reselect the subject.
4. Without deselecting the image, select the Shadow layer.
5. Press Option+Delete. This creates a solid black shape that exactly matches the shape and position of the selection and places it in the Shadow layer.
6. With the Shadow layer still selected, choose Edit > Transform > Distort.
7. Grab the dot in the middle of the top bar and drag down to create the shadow, as shown in **FIGURE 9.35**. Click OK or press Return.
8. With the Shadow layer still selected, change Opacity to 25%.
9. Choose Filter > Blur Gallery > Tilt/Shift to apply a variable blur.

FIGURE 9.35 The black shape, which matches the subject, is distorted to look like a shadow cast by the glass.

FIGURE 9.36 Grab a rotation dot (left arrow) and rotate the shadow until it is sharp near the glass and softer farther away.

10. Rotate the Tilt/Shift blur until the shadow closest to the glass is sharp, while that farthest away is blurry (**FIGURE 9.36**).

11. Adjust layer opacity to suit.

Done!

This illustrates how different aspects of Photoshop combine to create an effect. Most of the time, the effects we need to create are simple: add text, crop an image, adjust exposure, and so on. However, Photoshop is a very deep and powerful tool; the more you explore, the more there is to discover— and have fun with.

Photoshop is a very flexible, powerful and deep tool— the more you explore, the more there is to discover.

KEY POINTS

This chapter wraps this section on still images. In the next section, we take everything we've learned and apply it to moving images. And, as you might suspect, that starts with planning.

When it comes to this chapter, here's what I want you to remember:

- Composite images are images composed of multiple images or elements such as text.
- The Layers panel allows us to add, rearrange, modify, and delete elements.

- While the Background layer can be edited, converting it to a layer provides more flexibility, including changing position and transparency.
- Choosing Edit > Free Transform is a useful way to modify elements on a layer.
- To add multiple elements to the same image, choose File > Place, or drag the elements into the image from the desktop.
- Selections mark areas of a layer to be modified.
- There are many different selection tools, each designed for a specific purpose.
- Filters change the look of an image or selection.
- Blurring a background makes text easier to read.
- Blend modes combine textures between layers.

PRACTICE PERSUASION

Your challenge is to use "nonsense" text to create two ads using Photoshop. The purpose is to separate emotions from content. Then, see how composition and color combine to convey a message.

Create two Photoshop ads. One is for a boutique cupcake store, and the other is for a plumbing contractor. Each ad:

- Must use *only* the pseudo-Latin text from www.lipsum.com or the pseudo-Latin characters that travel with the Text tool.
- The actual content of your words must be gibberish.
- You can use images.
- Must have a Call to Action with a URL.

The goal is to get you thinking about design, not writing. Visuals, not content. Think about how a message *looks*, rather than what a message *says*.

Everything else is up to you.

RE-CREATING NORMAN ROCKWELL'S CREATIVE PROCESS

Loren Miller (and Rachel Victor)
Producers, Jarvis Rockwell

Jarvis Rockwell is a film I co-produced with Rachel Victor. (It was released in the summer of 2020.) It's a biography of Jarvis Rockwell, the oldest son of Norman Rockwell, one of America's most famous illustrators.

The core story of our film involves how Jarvis veered from his dad's expectations that he would also become a realist oil painter. Jarvis showed early brilliance as a sketch artist, with little formal training. But Jarvis was destined for other, more conceptual paths, as were his younger siblings. Tom became a novelist and poet, and Peter became a stone sculptor.

Tom and Peter both credit Jarvis for navigating difficulties with "Pop," chief among them his preoccupation with work. At the heart of that preoccupation was Norman's actual process for developing his iconic paintings of 20th century American life.

It wasn't just the content, but how he got there. Jarvis paints his father as a workaholic "eight days a week." Norman's process very much resembled the previsualization planning and techniques of a contemporary filmmaker. Norman was brilliant. But the process took a toll on him and his family. He never called himself a painter or artist; he used the term "illustrator."

Norman's art editors at the *Saturday Evening Post* needed his iconic cover paintings on time, demanding a full-sized canvas every other month, and he had to find a method to accelerate the process to fulfill the images in his head and meet nearly impossible deadlines. He successfully delivered 323 classic cover paintings for the *Post* over 47 years.

When Jarvis visited his father's studio, Pop was usually preoccupied with what Jarvis perceived as an unreal world of props and costumes, awkward poses, cameras, lights, and photo projectors, or actual paintings.

Norman would draw him into the process, using him as a child model. This role-play often substituted for family time and became the heart of Jarvis' identity crisis, which has taken a lifetime to shake off.

By showing the detailed process Norman used to construct his illustrations, we highlight the extent to which Norman was preoccupied with his elaborate workflow, resulting in a mutually challenging environment for father and son. This workflow, which allowed Norman to provide for his family and brought him fame, resulted in Jarvis' adult struggle to forge a different, ultimately successful, path.

Norman directing a model with Louie Lamond at camera (left). Printed reference photo of young Jarvis for "Homecoming Marine" (right). (Courtesy of Rockwell Family Agency, Norman Rockwell Museum Image Services.)

First, Norman would freehand-sketch a previsualization of a proposed painting. It would contain all the characters he envisioned and a basic setting such as a luncheonette, a garage, a barbershop, or a train car.

Then he would cast photo models—often his sons, as well as his wife, neighbors, townspeople, even himself.

Then he would painstakingly direct the poses and expressions of his models. He has stated, "Directing models is an art unto itself."

Employing a cycling cadre of professional cameramen, he would photograph several takes of each model with a Speed Graphic 5"× 7" plate camera, the high-resolution model of its day.

He would develop these studio reference photos and print them himself in a darkroom alongside the studio. (Today you can visit the Digital Collection at the Rockwell Museum's website to view them.)

He would lay out the prints of each model on a table or window seat to select the best takes of each, searching for the most effective facial and body expressions.

Process hardware of the day: Norman's Speed Graphic plate camera (left). Museum Image Services' Thomas Mesquita handles Norman's Balopticon projector for our insert shooting as producer Loren Miller looks on (right). (Image Credits: Courtesy of Harlan Reiniger. © 2019 Rachel Victor Films, LLC (left). Courtesy of Loren Miller. © 2020 Rachel Victor Films, LLC (right))

Then he would cut and snip each character into a highly composed photo collage.

These "edited selects" replaced large parts of the original previsualization sketch to such a point that few if any exist today.

He would then photograph the resulting collage itself and load the resulting print into a Bausch and Lomb Balopticon projector, aimed at his full-size canvas.

Re-creating Norman rotoscoping Jarvis onto canvas from projector. (Photo credits: Loren Miller. © 2020 Rachel Victor Films, LLC.)

He would then trace each face and body in fine detail onto the canvas using a charcoal pencil as shown in the image.

Finally, he would mix the paints and leverage his Art Student League training and expertise in oils. If you compare his photos to the final paintings, you often see exact expressions preserved, but with large amounts of invented reality; it was not at all a "paint by numbers" process. But the painting would be done in record time and shipped to the *Post*.

Norman never kept the process a secret—he even composed a booklet for art students detailing it. Many artists use it.

Our goal, with the images we've created is to bring our audience to better understand Norman Rockwell, the artist, and Jarvis Rockwell, his son, whose journey has been an amazing path no less unique than his father's and well worth the trip.

SECTION 3

PERSUASIVE MOVING IMAGES

SECTION GOALS

As we move from still images into video, everything we learned about still images also applies to video with one big exception: In video, everything moves! Talent moves, cameras move, titles and graphics move. Because of this movement, our persuasive options expand.

In this section, we explore just what all this movement means; plus, two chapters on interviews and audio that apply to most video projects. The chapters in this section are:

- **Chapter 10: Video Pre-production.** How to plan a video shoot and manage your media.

- **Chapter 11: Create Compelling Content with Interviews.** How to improve your interviewing techniques.

- **Chapter 12: Sound Improves the Picture.** An audio overview from gear to recording and editing.

- **Chapter 13: Video Production.** How to shoot effective video.

- **Chapter 14: Video Post-production.** An editing workflow from import to final output.

- **Chapter 15: Motion Graphics: Make Things Move.** An introduction to motion graphic videos.

- **Chapter 16: Advanced Motion: Particles, Paths, and Perspective.** Advanced and exciting motion graphic techniques. This chapter, and several tables of keyboard shortcuts, are only available online. Go to: www.peachpit.com/visualpersuasion to register your book and download this chapter.

FOUR FUNDAMENTAL THEMES

- Persuasion is a choice we ask each individual viewer to make.

- To deliver our message, we must first attract and hold a viewer's attention.

- For greatest effect, a persuasive message must contain a cogent story, delivered with emotion, targeted at a specific audience, and end with a clear Call to Action.

- The *Six Priorities* and *the Rule of Thirds* provide guidelines we can use to effectively capture and retain the eye of the viewer.

THE SIX PRIORITIES

These determine where the eye looks first in an image:

1. Movement
2. Focus
3. Difference
4. Brighter
5. Bigger
6. In front

My biggest pet peeve is when a producer is not organized and ends up wasting people's time. Shots need to be planned, equipment ready, etc., so the day can go as smoothly as possible.[1]

ALLISON WILLIAMS
ACTOR

CHAPTER 10

VIDEO PRE-PRODUCTION

CHAPTER GOALS

Video and its cousin, motion graphics, are all about movement—movement through characters, through emotions, through time, through space, and, most importantly, through a frame. This chapter extends what we've learned so far into video. You'll discover that while there's a direct correlation between still and moving images in terms of how they are created, video creates a *whole lot more* images that we need to work with.

The goals for this chapter are to:

- Plan a video shoot
- Create storyboards and refine a video workflow
- Define basic video terms and concepts
- Explain storage, media, and media management

[1] Personal email.

If we don't first get someone's attention, the rest of our work is lost.

MOVEMENT IS THE DIFFERENCE between photographs and video. This statement is both obvious and deep. What does it mean to say, "Everything moves"?

Our goal, as we learned in Chapter 1, "The Power of Persuasion," is to control the eye of the viewer, and nothing captures the eye like movement. For tens of thousands of years, humans were hunters—or were hunted. If something moved, our first thought was to see if we could eat *it* or if it was likely to eat *us*. This deeply instinctual reaction is why movement attracts our attention; it's a self-defense mechanism that we can no more turn off than we can stop breathing. It happens automatically.

Why is this important? Because if our primary goal is persuading someone to do something, then we must first get their attention. If they never see our message, all the rest of our work is lost. I can't stress this enough: If we don't first get someone's attention, the rest of our work is lost.

The problem is that, in today's distracted, stressful, self-absorbed world, getting someone's attention is increasingly difficult. We watch videos with the sound turned off, listen to audio podcasts that have no images, and glaze over web ads without ever registering their content.

Your biggest weapons in combatting this distraction are movement for video, sound effects for audio, and alluring images for stills. Movement is, by far, the most important. Movement attracts the eye, while sound effects attract the imagination.

There are two ways we can create video: We can record it with a camera, as we do with film or video, or we can create it using the computer, as we do with motion graphics and animation. In this chapter, I want to focus on pre-production, the process of planning a video, and discuss ideas that are common to both motion graphics and video.

Two Key Camera Concepts

The camera represents the point of view of the audience.

Camera work does not need to be perfect to be effective, but it does need to be good.

Two Key Camera Concepts

To start, let's reinforce two key concepts for video. First, the camera represents the point of view of the audience. This means that whenever you move the camera, you are grabbing the audience by the collar and dragging them with you.

Second, camera work does not need to be perfect to be effective, but it does need to be good enough to retain an audience. In fact, there's a whole counter-aesthetic of creating intentionally "rough" camera work to make the video seem more believable. (Think of the *Jason Bourne* films directed by Paul Greenglass.) However, taken to extremes, shaky handheld camera shots make a video look amateurish. Like all creative options, there's nothing inherently wrong about handheld shots, but be careful. Move too quickly, or too abruptly, and your audience will get motion sick, or so distracted by all the excessive movement that they stop watching. Gently, please. Simply because something looks amateurish does not make it better or more "true."

Pre-production is the process of planning a video.

Feel the Rhythm

An underlying theme when working with video is "rhythm." Coherent movement is rhythmic. Chaos is not. Coherent speech has a rhythm, separate from the oratory of professional speakers. Each of us, as we speak, gesture, or move, does so rhythmically. Whether this rhythm is determined by our heartbeat or breathing or the fact we have opposing legs and arms, I am not really sure. What I am sure about, though, is that this rhythm exists everywhere, not just in music or poetry, but in interviews, production, and, especially, editing.

As you will see in the "Practice Persuasion" exercise at the end of this chapter, this rhythm is easy to spot once you start looking for it. We talked about this in Chapter 3, "Persuasive Writing," when we discussed the power of three—"On your mark, get set, go!" But rhythm is more pervasive than that. When two people shake hands, the timing of how they raise their hands, shake, then drop them again is almost dance-like. When two friends walk together, they tend to be in step. When someone is finishing a persuasive talk, they punctuate their closing points rhythmically.

Just as we looked at the deeper, emotional context of words, fonts, colors, and images, the process of recording and editing moving images also harnesses the power of rhythm.

Copyright Applies to Video, Too!

We talked about copyright in Chapter 7, "Persuasive Photos." These days copyright extends to more than words. It includes graphics, logos, photographs, the way people look, the colors objects are painted—copyright extends across a *wide* variety of creative work. If you plan to start creating and posting media, spend time reading up on copyright. Even better, talk with a lawyer about what you want to do. Take-down notices, copyright infringement, lawsuits—and their resulting costs— can be easily prevented if you spend time up front learning the rules of the road.

Planning a Video

Above all, when planning any video, remember that nothing *ever goes according to plan.*

Whether you are creating a video (Chapter 13, "Video Production") or a motion graphic (Chapter 15, "Motion Graphics: Make Things Move"), it all starts with a plan. While the elements vary between videos created in real life or the computer, the underlying concepts are the same.

Above all, when planning any video, remember that *nothing* ever goes according to plan. "You have to understand one thing about directing," says film director Ron Howard. "Every project you get involved with, ultimately, is just going to find its way to breaking your heart."

Sigh...I've lived that advice.

Planning encompasses a vast array of tasks:

- Developing the concept, message, story, and script
- Logistics, such as locations, gear, permissions, and so on
- Building the team, including talent, crew, and support staff

All of these are critically important. You need to plan, schedule, and collect all the things that need to be delivered to the same place in time for the shoot. These are important but outside the scope of this book.

Equally important, though harder, I think, is the creative planning: the look, shots, staging, blocking, and framing of your images. This is harder because there are so many options. A musical octave has only 12 notes, but, oh, the music you can make with it! It's the same with images; the creative options of video are so limitless that it is easy to get paralyzed with indecision.

Storyboards Are a Tool for Thinking

A useful tool to help you plan is a storyboard. Storyboards are not art, which is readily apparent in the storyboard I drew for FIGURE 10.1. Rather, consider storyboards as "thought sketches" to help you think about your project visually.

A quick web search will turn up lots of storyboard creation tools, including some featuring beautiful, professionally drawn artwork. But, don't get distracted. A successful storyboard does not have to look great; it just has to help you think about what you want the audience to see.

For example, the template used in Figure 10.1 has two boxes stacked in a column. In the top box I sketch the image I want the audience to see, in other words, the point of view of the camera. Then, in the lower box, I draw

FIGURE 10.1 A storyboard is a tool to help you think visually. The top square in each column shows what the audience sees; the bottom is a floor plan showing positions and movement to create that shot.

a "floor plan" illustrating the position and movement of talent, cameras, set pieces, and so on. (I add a dialogue reference just so I can tie the image back to the script.)

These sketches allow me to communicate quickly with the crew and talent about the type of shot, where gear can be placed, the location of the camera, and so on. By first sketching my ideas, everyone on the team can discuss a shoot before production even starts. Then, during production, everyone knows where we are and where we are going.

For a sit-down interview, storyboards won't be very helpful. But as soon as cameras or talent start moving, you start using multiple sets, you start creating a motion graphic, or the crew gets larger than a couple of friends helping out for a few hours, the time you take to sketch a storyboard will speed communication on set.

Consider storyboards as "thought sketches" to help you think about your project visually.

A Workflow Is Essential

Françoise Bonnot won an Oscar for editing *Z* by Costa-Gavras and a BAFTA for editing *Missing*. In 2003, I met her for one brief conversation, which you'll read more about at the end of Chapter 14, "Video Post-production." During our 15-minute chat, she made a comment that changed how I teach, how I write, even how I approach projects. Her comment was, "I never have enough time to finish a project to my satisfaction." And she's won an Oscar for her work! If she doesn't have enough time, then we certainly don't.

A workflow is an explicit, usually written, series of steps (a process) for completing a task efficiently.

This is why I'm so focused on efficiency and workflow. There is *never* enough time. Lighting forgot to bring extension cords for the lights. Makeup needs more time for hair. The teleprompter breaks. As the famous sage Roseanne Roseannadanna said, "It's always something." There is barely enough time to do something once; there is no time to do it again.

This is why planning, preparation, and workflow are so important. Since we don't have enough time, we need to allocate it as efficiently as possible. The more time we spend up front planning, visualizing, and thinking, the more efficient we are when the time comes to execute our vision in production and post.

With that as background, here's a simple workflow for any creative project:

1. **Plan.** Define your goals, budget, audience, deliverables, message, and story. Oh, and determine the deadline.

2. **Gather.** Collect the elements you need to complete the project.

3. **Create.** Create your project.

4. **Critique.** This is the hardest task: seeing what you've actually created, then making it better. This can often be helped by asking the opinion of others.

5. **Revise.** Improve your work based on these critiques. Then, go back and critique again. Repeat until the deadline arrives.

6. **Release.** Share your work with the world. If your planning was successful and you stayed on track, your project will be ready.

For instance, there's no value to creating a video if you don't know what you need to deliver, your deadlines, or your audience. All you are doing is spinning your wheels.

Defining Basic Video Terms

Like most industries, there is a lot of jargon in media. It sometimes seems that we make our language unnecessarily complex just so those of us working in media feel, uh, "special." (Not content to call a wooden clothespin a "clothespin," film crews for decades have called them "C-47s." The reason is amusing; Google it.) Not to worry, I won't define every term, but here is a baker's dozen that I'll use for the rest of this book.

Codec. Short for "COder/DECoder," a codec is a software-based algorithm that converts the reality of light and sound into binary digits that can be stored on a computer, then played back. There are hundreds of codecs running loose in the wild! Codecs are optimized for high image quality, small file size, or ease of editing. For technical reasons, no codec provides all three, as **FIGURE 10.2** illustrates.

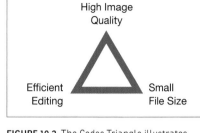

FIGURE 10.2 The Codec Triangle illustrates why we care about codecs: When working with media, we can choose only two of these three options.

Deliverable. This includes the content and technical specifications of the master file you need to create for distribution to your audience. If you are posting a video to YouTube, the tech specs are pretty loose. If you are sending a file to Netflix, its specs are really strict. It is wise to verify exactly the specs you need to deliver before you start shooting.

Clip. A video clip is a sequence of still images, called "frames," that flash on-screen quickly (**FIGURE 10.3**). A video clip can contain video frames, audio, or both. Clips may also contain data such as timecode, captions, and labels (metadata).

Frame. (1) A frame is a single image in a video clip, or (2) the boundaries of an image.

Framing. Framing is how the elements inside an image are positioned relative to the frame.

Frame size. The size of a frame is measured in pixels. A common video frame size is 1920×1080 pixels, with the horizontal dimension always listed first.

When it comes to shooting and editing video, we generally use a limited set of video codecs (though there are hundreds of codecs to choose from):

- H.264 and its cousins, AVCHD and AVCCAM
- HEVC
- MPEG 2, 3, or 4
- Apple ProRes
- Professional formats like RAW, Log-C and P2

FIGURE 10.3 This is the first-ever movie, recorded in 1878 by Eadweard Muybridge. It illustrates that film and video are a series of still images played quickly to create the illusion of motion. ('The Horse in Motion' Photographed by Eadweard Muybridge, Copyright 1878)

Frame rate. This determines the number of individual frames in a clip displayed each second. Typical frame rates include 24, 25, 29.97, 30, 48, 50, 59.94, and 60 frames per second (fps). It is *really* helpful to shoot using the frame rate you need to deliver.

Aspect ratio. This is the ratio of the frame width to its height (**FIGURE 10.4**). Most high-definition (HD) video has an aspect ratio of 16:9. Older standard-definition (SD) video used a 4:3 aspect ratio.

FIGURE 10.4 Aspect ratio. The wider version, 16:9, is used in movies on computers and TVs today. 4:3 was what we watched for the first 50 years of TV.

Duration. The duration is the length, in hours, minutes, seconds, and frames of a video clip, or a portion of a video clip, called a "range."

Timecode. Timecode is an "address" that uniquely labels each frame in a video clip. It is expressed as a time value: *Hours:Minutes:Seconds:Frames*. For example, a timecode location of 01:24:48:10 translates to 1 hour, 24 minutes, 48 seconds, and 10 frames. There is no necessary relationship between time of day and timecode.

iPhone 7 and later have two video settings: High Efficiency (HEVC) and Most Compatible (H.264). This setting determines which codec the phone will use for recording. Both are highly compressed, though HEVC requires a fairly new computer and operating system for smooth playback and editing. I recommend using H.264, though file sizes are larger.

Transcode. This means to convert from one codec to another, for example, to convert from H.264 to ProRes. Generally, transcoding does not change the frame size or frame rate, though it could if needed.

Compress. (1) Compress means to remove data to reduce the size of an audio or video file. We always compress master files before posting them to the web because the file sizes are so big. (2) It can also mean to adjust audio levels to reduce the amount of variation between the loudest and softest passages. Audio compression does not remove data. I'll cover audio compression in Chapter 12, "Sound Improves the Picture," and data compression in Chapter 14, "Video Post-production."

Camera master. This is the original media file as recorded by the camera. (This is also called "camera-native" media.) Many camera masters are transcoded before they are edited.

Media Management

Media management is the process of figuring out how to handle your media before you start creating it. It is much wiser to spend time planning where you will store, how you will transport, and how you will name media files before you start than to scramble at the last minute to figure out where a missing file went.

The reason media management is such a challenge for media projects is that, unlike a word processor or Photoshop document, your media elements are not contained in the media project itself. Rather, they are "linked," or pointed to, from within the project to wherever they are stored elsewhere on your system.

This means is that if you move a media file or a folder, rename a file or folder, or, heaven forbid, delete a file or folder from the desktop, all the links inside your project break. Broken links mean you can't export a finished project; in fact, you can't even *play* your project.

I can't count the dozens of emails I've gotten over the years from panicked editors who, just before outputting their finished work, decided to "clean things up a bit" and renamed or moved files only to discover the catastrophe that caused.

It is much—much!—wiser to get all your files organized before you start editing. This way you know where everything is before you start, and you never have to worry about whether you can export your project when the time comes to finish your work.

Another reason to think about your media before you start a project is that media files are huge! It is not unusual for even a small project to contain dozens of gigabytes of data. I have several projects that contain terabytes of files.

Like dandelions in the spring, media files seem to multiply and take over your system.

Media management helps us plan how to label, store, find, and access media files during production and editing.

Fast, Big Storage Is Essential

It is impossible to describe just how big media files are until you start to store them. Media management helps us plan how to label, store, find, and access media files during production and editing. That, as many editors have discovered, is much easier to say than to do.

For example, standard-definition (SD) video, which your parents and grandparents happily watched on TV, holds 345,600 pixels per frame and takes about 13 GB to store an hour-long video. A 4K frame, which we watch today on Netflix, holds 8.3 million pixels per frame and, depending upon the codec and compression, could require more than 400 GB to store an hour! (Most media projects involve dozens to hundreds of hours of raw material!)

FIGURE 10.5 illustrates the trend toward shooting larger and larger frame sizes, which creates larger and larger file sizes, which requires more and more storage space. Worse, since digital files are so easy to create, not only are digital files getting bigger, we tend to create more of them than we need for each project.

FIGURE 10.5 A comparison of video frame sizes from SD to 6K. The newest cameras now shoot 8K and larger frame sizes, which won't fit on this page.

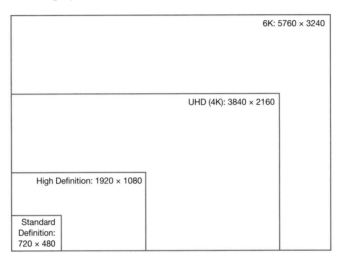

6K: 5760 × 3240

UHD (4K): 3840 × 2160

High Definition: 1920 × 1080

Standard Definition: 720 × 480

Frame Sizes Continue to Grow

A 6K video frame, which is used in production and video editing (but not yet for final distribution), holds around 19 million pixels *per frame* and, depending upon the codec, requires about 1 TB to store one hour of media. 8K frames are twice as big as 6K! Sigh....

As an example, **TABLE 10.1** illustrates how the file size of iPhone video varies by frame size, frame rate, and duration.

What's important in Table 10.1 is not any specific number but that video files are measured in *dozens to hundreds* of gigabytes. Figuring out where to safely store this much media becomes important.

The bigger lessons of media management are:

- Media files take *huge* amounts of space—so much so that if you intend to get into media production or editing, you will need to invest in additional storage to expand that which is built into your computer.

- Media files also need fast storage. Media transfer rates, the speed that data needs to travel from your storage to your computer to provide smooth playback, generally require dozens to hundreds of megabytes per *second*.

- As frame sizes increase, storage capacity and speed requirements increase dramatically.

- When you start editing, you need to reserve roughly four to eight times more storage space than your original camera master files. This allows for transcoding and work files generated by all editing systems.

TABLE 10.1 iPhone Video File Sizes

	1 MINUTE H.264	1 HOUR H.264	1 MINUTE HEVC	1 HOUR HEVC
720p HD @ 30 frames/sec	60 MB	3.5 GB	40 MB	2.4 GB
1080p HD @ 30 frames/sec	130 MB	7.8 GB	60 MB	3.6 GB
1080p HD @ 60 frames/sec	175 MB	10.5 GB	90 MB	5.4 GB
1080p HD slo-mo @ 120 frames/sec	350 MB	21 GB	170 MB	10.2 GB
1080p HD slo-mo @ 240 frames/sec	N/A	N/A	480 MB	28.8 GB
4K HD @ 24 frames/sec	270 MB	16.2 GB	135 MB	8.1 GB
4K HD @ 30 frames/sec	350 MB	21 GB	170 MB	10.2 GB
4K HD @ 60 frames/sec	N/A	N/A	400 MB	24 GB

Most video files are far too large to email. HEVC recording is available only on iPhone 7 or newer.

TABLE 10.2 converts between video formats shot by a phone and formats commonly used in editing. Again, look at the size of the files.

TABLE 10.2 Video File Sizes by Frame Size and Codec

SOURCE FILE	H.264 1 HOUR	PRORES 422 1 HOUR
720p HD @ 30 frames/sec	3.6 GB	33 GB
1080p HD @ 30 frames/sec	7.8 GB	66 GB
4K UHD @ 30 frames/sec	21 GB	265 GB

There's no such thing as "too much storage."

This reinforces the importance of thinking about where you'll store your media once it's shot. If extensive media recording or editing is in your future, getting enough storage is as important as having a fast computer to edit it.

Cloud Storage

For mobile device users, iCloud and Google Drive, among others, are options for storing media, depending upon the plan you have and the amount of media you have to store. Either is also a good way to move files from one device to another. However, the Internet connection between your local computer and the cloud is generally not fast enough for editing. So, while you can use the cloud for backups and transfers, you will still need to copy your media files locally for editing. This also means you need to allow time for those files to transfer, both up and down.

While there are a number of new technologies that allow editing media in the cloud—Bebop Technology, Blackbird, and Premiere Rush being three examples—I still recommend editing media locally for now.

There's Never Enough Storage

Media editors obsess about their computers, including whether they have enough RAM or a fast GPU or whether they should get a 3.1 or 3.2 GHz CPU. What they don't realize is that the speed and capacity of their storage will have a greater impact on performance for editing than the speed of their computer.

A fast computer is great, but a fast computer with slow storage or, worse, no extra storage at all won't do you a whole lot of good. Storage technology changes hourly, so providing suggestions for specific gear in a book gets very dated very quickly. Still, there are general guidelines that are useful in planning storage needs.

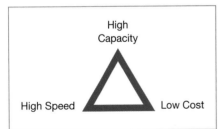

FIGURE 10.6 The Storage Triangle, like the Codec Triangle, illustrates that you can optimize storage for only two criteria.

Like codecs, we also have a selection triangle for storage (**FIGURE 10.6**). Storage can have high capacity, high speed, or low cost, but you can optimize for only two.

- Solid-state drives are blindingly fast and affordable, but they don't hold a lot.
- Single hard drives hold a lot, are low cost, but aren't fast.
- RAIDs hold a lot, are fast, but are not cheap.

Before you buy any storage, ask yourself these questions:

- What codecs am I using? Storage needs to be at least twice as fast as the normal data rate of your media.

- How much do I plan to shoot? Every production always shoots more than they expect. Plan to store two times your planned shooting ratio.

- How do I plan to edit? Editing generates *lots* of work files. Plan to store four to eight times the space needed for the media that you shoot for these extra files.

There are no hard-and-fast rules. In my office, I have more than 90 TB of storage, which is a lot. It's also almost full. There's no such thing as "too much storage."

A fast computer is great, but a fast computer with slow storage or, worse, no extra storage at all won't do you a whole lot of good.

Backups Are Essential

It's been said before, and it is worth repeating: Back up your media!

Yes, backups take space. Yes, storage costs money. No, you haven't had problems before. But, can you afford the risk of losing all your media? I can't. Whether you back up to a second hard drive, copy files to iCloud or BackBlaze, or transfer them to a local server, always make sure you have at least two copies of your media. Backups are far less expensive than reshooting.

Backups should include all media, work files, and projects.

Track Your Media

When we create a new word processing document, we type what we need, give the file a name, save it, and move on to the next task. By now, you should suspect that media files aren't that simple. You'd be right.

These days, lower-cost cameras record media in chunks, which are stored in different folders on the camera card. (Mobile phones consolidate these files before you transfer them to your editing system.) If you are transferring files directly from your mobile device to your computer, transfer speeds matter. For example, connecting via a cable or using WiFi or AirDrop will be faster than copying files to the Internet and back.

If you are copying a camera card to your hard disk, first create a folder on your hard disk and name it something meaningful. Then copy the *entire* contents of the camera card—every folder—into that new folder on your hard disk. Many camera formats store data across multiple folders; if you don't copy all of them, something in your media may break.

Don't rename the camera master files or the folder names from the camera card; use the names given them by the camera. This allows all these different media files to successfully import into your editing system.

Rather, rename the folders that *you* created on your storage that you copied the media card data into. Every editing system (called an "NLE" for "non-linear editing," such as Apple Final Cut Pro X, iMovie, Adobe Premiere Pro, Premiere Rush, Avid Media Composer, and the rest) allows you to rename clips inside the application.

A Folder Naming Convention

While naming conventions can and should vary depending upon your project, **FIGURE 10.7** illustrates one that I like and recommend when using camera cards.

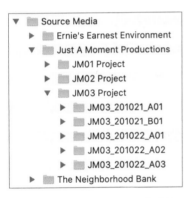

FIGURE 10.7 An easy-to-use media folder naming system.

First, create a folder on your hard drive in which to store all your media. Putting everything in one place makes it easier to find and back up. Inside this master media folder, which I call "Source Media" in Figure 10.7, I create a folder for each client, campaign, or theme; for example, one client is named "Just A Moment Productions."

Inside each client folder, I create a folder for each project. I use a two-letter code for these, as in "JM01." *JM* stands for the name of the client, and *01* stands for the first project I created for that client.

Inside the project folder, I create one folder for every camera card or, if I'm transferring files from an iOS device, a folder that contains the smartphone movies. Unlike media stored in a camera card, smartphone movies are self-contained.

The names of these camera card folders start with the client code, then the date the media was shot, followed by the camera number and the number of cards shot that day. For example, "JM03_201022_A02" translates to: the third project for Just A Moment Productions, shot on October 22, 2020, on the A (primary) camera. It was the second camera card shot that day.

> ### Think Before You Start
>
> With all this talk of naming and storing, you should be getting the idea that it makes sense to think about how you want to name your files and where you want to store them before you start a new project. Experience teaches that solving these issues *before* you start shooting and editing saves a lot of anguish when you know you shot something but can't find it.

You are welcome to modify this naming convention as you see fit. However, keep the following media concepts in mind as you do:

- Folder names that you copy media into can be named whatever makes sense to you.
- Don't change the names of any files or folders that were created by the camera and stored on a camera card.
- Copy the *entire* contents of the camera card into its own folder, one folder per camera card.
- Media shot on mobile devices should be copied into their own folders, then grouped as you see fit. Unlike camera card media, smartphone files can be renamed in the Finder.
- Rename files inside your editing software, using names that make sense to you.
- Allow four to eight times the space needed to store your camera master files for editing work files.
- Always back up all media and project files. Always.

If you can't find the clip you need, it's the same as if you never shot it.

KEY POINTS

We've covered a lot in this chapter; here's what I want you to remember:

- A key difference between video and other creative work is that elements in video move and video files are huge.
- All human activity, including speech, is rhythmic.
- Planning a video shoot includes logistical and creative decisions.
- There is never enough time to work on any production; planning and workflow are essential.
- Video includes a lot of technical terms. Understanding what they mean improves communication between team members.
- Before you start any project, make sure you understand the technical and content specifications of what you need to deliver and when it's due.
- Give yourself plenty of space to store your media.
- Plan where to store media, how it will be organized, and how it will be named before starting editing.
- If you can't find the clip you need, it's the same as if you never shot it.

PRACTICE PERSUASION

Storyboard the first five shots (after opening titles) in a favorite movie. The goal here is to get you to think about where elements are located: camera, talent, and set. Which way is the camera pointing?

Try to envision what the actual storyboard for that movie would look like. I'm less interested in what's in the camera's view (top box) and more interested in the floor plan you sketch in the lower box that shows where everything is located.

PERSUASION P-O-V

THE IMPORTANCE OF VIDEO SKILLS FOR STUDENTS

Vicki Aleck
Chief Engineer
onfocusvideo.weebly.com

McCormick, South Carolina, was a thriving town, surrounded by cotton plantations in the late 1800s and early 1900s. Then, in 1922, the boll weevil came and destroyed all the crops. Since then McCormick has struggled, at one point being the poorest county in South Carolina. Many graduates of the high school stay within a 20-mile radius of the city all their lives.

The city had a poor rural school system, but they did have access to grant funds if there was a career implementation program. I moved there from Chicago, having worked at WLS-TV, Chicago, for more than 20 years. Working for a network-owned and operated station, I had skills that could be shared with the schools here.

The use of moving pictures, along with complementary skills in audio and lighting, remains an important skill. If a broadcast studio could be built at the school complex, we could open students' eyes to the opportunities in media and video production. The program could also include speakers from nearby cities, along with tours of their facilities. This way, students could see how the skills they learn at school can transfer to their careers after graduation.

A broadcast studio touches so many skillsets. There are hard skills: how to operate the equipment, cameras, switcher, streamer, lighting, and audio. There are also the soft skills: writing, interview techniques, booking guests, creating the show rundown, and so on.

We were fortunate to get a grant and start building. We chose equipment that will be viable for years such as PTZ production cameras with HDMI connectors for the studio, operated by an IP camera controller, which is a common configuration at the major TV networks. Each camera can stream independently of the switcher, which is a plus for live events like a County Council meeting.

The switcher has virtual sets, includes HDMI inputs, and streams digital media. It's also connected to the National Weather System. Program times can be scheduled through the switcher to run anytime on screens throughout the school. Our control room doubles as a sound booth for voice overs and podcasts. We have a few Sure Shot–style HD cameras and wired lavalier microphones for field work.

We've barely scratched the surface in terms of the ways we can use this equipment. Everyone has ideas, and our aim is to make their ideas a reality. It is my hope and my goal that exposure to these skills will change and guide our students' lives.

Well, the first thing I tell anybody who's going to be doing interviews is homework. I do so much homework, I know more about the person than he or she does about himself.[1]

BARBARA WALTERS
JOURNALIST

CHAPTER 11

CREATE COMPELLING CONTENT WITH INTERVIEWS

CHAPTER GOALS

We can create our own persuasive stories, or we can discover them in others using interviews. This chapter looks at the process of interviewing, both as a host and as a guest. Key points include:

- How to plan an interview
- How to conduct an interview
- How to be a good interview guest

An interview is a formalized process of sharing information with an audience within a very limited time.

THERE ARE TWO WAYS to create persuasive stories: We can write them ourselves, or we can learn them from others. In Chapter 3, "Persuasive Writing," we looked at better ways to tell stories by creating the content ourselves. In this chapter, I want to discuss how to discover compelling content through interviews.

Interviews are a staple of all media, because they work really well at revealing information. I've conducted thousands of interviews for broadcast, news, live events, and my podcast, DigitalProductionBuzz.com. The majority were conducted over the phone or Skype. I've always enjoyed interviewing people, because every time I do, I learn something new.

Interviews can be video or audio; it doesn't really matter, because the content is more important than the visuals. Especially in today's world of audio podcasts, interviews are a fast and easy way to learn something new. Interviews are conversations: two voices expanding a subject into our imagination.

An interview is a formalized process of sharing information with an audience within a limited time. We are all familiar with a heart-tugging interview, where the emotions and passion of a subject are reflected in the face and voice of the person speaking. But interviews are much more than that. Their purpose is to gain information from a subject-matter expert that you can't get any other way. Interviews also provide emotional insight into a topic that is often impossible to capture through narration or images alone.

Planning the Interview

Your role, as the interviewer, is to represent the audience, to ask the questions they would ask the guest if they were there. As the interviewer, you are in charge. I often describe it as a dance where the interviewer is leading and the guest is following. They both need to work together for the interview to work. Interviews are *controlled* events: They will be controlled by either the interviewer or the guest. Unless you want the guest to take charge, you need to plan what you want to learn and where you want the conversation to go (**FIGURE 11.1**).

The first step is to think about your goal. What do you want the interview to accomplish, what key points do you want to cover, and what do

FIGURE 11.1 All interviews start with a plan of what you want to accomplish. (Image Credit: pexels.com)

you want the audience to learn? The old newspaper questions of who, what, when, where, why, and how remain relevant. In fact, I use them as the structure to all my interviews.

An interview, in most cases, is not an interrogation. You are not *60 Minutes*. Rather, you and the guest are having a conversation about a subject you are both interested in. The more comfortable you make the guest, the better answers you are likely to get.

Questions should encourage the subject to answer in some detail, but not to embarrass them or pin them to the wall. Be respectful of their position: Some questions may be inappropriate given their level within the corporation or organization. I once asked a technical engineer on a live show why his company filed bankruptcy the week before. It was an appropriate question given the situation, but he was the wrong person to answer it. Have a list of questions, but be ready to move into different areas if the discussion seems interesting—you can always return to your list of questions later.

Interviews are controlled events: They will be controlled by either the interviewer or the guest.

Planning is not as sexy as production, but it is just as essential. When time is limited—and when is time *not* limited during production?—you can't afford to wander around trying to discover your subject during an interview. You need to have a goal in mind. This doesn't mean you can't explore subjects that arise during the conversation, but the best interviews are ones where the interviewer has a clear idea of where the conversation is headed (**FIGURE 11.2**).

FIGURE 11.2 Interviews can be as simple as sitting around a table.

In my mind, there's a difference between planning and scripting. I plan, but never script; that is, I have my notes, but I never tell the guest what to say. If I am interviewing an artist, I have no idea what they are going to say. But I do have an idea about whether my interview is going to discuss her life, her work, her achievements, or a cause she's pushing. Because most of my interviews are only a few minutes long, I need to focus my questions to get to the meat of the matter as quickly as possible. Asking focused questions also keeps the guest focused.

I always prepare questions in advance, one question for each minute I expect to spend recording the interview. While many of my interviews have been live, **FIGURE 11.3**, still more were recorded. Recording an interview provides greater opportunity to remove stumbles and get rid of unnecessary comments. All my recorded interviews run 20 minutes or less, which forces me to concentrate on getting the material I need, rather than go on a "fishing expedition." (This grew out of my on-location work in the early days of portable video tape. The tape cartridges held only 20 minutes of media.)

FIGURE 11.3 Interviews also can be as complex as live events, like the Digital Production Buzz 32-hour coverage of the 2017 NAB Show. (Image Credit: Digital Production Buzz / Thalo LLC)

Asking focused questions keeps the guest focused.

It Takes Two

Many video crews today are one-person operations, where the person asking the questions is also running the camera, doing lighting, and controlling audio. My recommendation, where at all possible, is to have a crew of two: one person running the gear and the other paying attention to the guest and content. It is hard to concentrate on content when you are constantly worrying about whether your camera is in focus or whether the audio sounds clean. An extra hand with the tech makes a big difference in the quality of your results.

Conducting an Interview

Once the planning is complete, it's time to conduct the interview, **FIGURE 11.4**. From my perspective, the interview "starts" the instant the guest appears. As the guest enters, direct your full attention to them. Introduce the crew and then let the crew focus on getting final tweaks made. As the interviewer, you need to build a relationship, a rapport, with the guest from the moment they walk in.

Virtually every guest, no matter how experienced, is nervous before the interview starts. They worry about how they look, what they'll say, and how you'll edit them. There's also the intimidation factor of having a camera stare at them. So, your first job is to decrease their fear.

FIGURE 11.4 Larry Jordan interviewing filmmaker Cirina Catania live at NAB 2018. (Image Credit: Digital Production Buzz / Thalo LLC)

Explain the general process of the interview, tell them where to sit, specify where to look, and reassure them that they won't die. (Don't laugh. You'd be surprised at the looks of relief I get when I tell a guest, "And, don't worry. I haven't had anyone die on camera... this week." They laugh, but they also relax.)

Asking questions is part art and part science. The art is in really listening to what your guest is saying. Actors call this being "in the moment," focusing intently on your guest and their comments. The science is in how you construct your questions.

I rarely ask my questions in the order I wrote them, and I often ask questions that aren't written because they spring from our discussion. In other words, the written questions act as a crutch when I need it, but they supplement, not replace, active listening.

Silence is not a bad thing, at least during a recorded interview. Sometimes, waiting to ask your next question gives the guest time to think about their last answer and then, after a pause, elaborate on it.

Your *core* goal for any interview is to find the story within the information the guest has to share and then get the audience to care about that story. To reach that goal, all my interviews have a definite structure.

> *Your role, as the interviewer, is to represent the audience and ask the questions they would ask if they were there.*

Keep Your Questions to Yourself

I never share questions with the guest in advance, because then the guest will try to memorize answers. Memorizing always sounds terrible! I *will* share general subject areas, but never questions. Also, if a guest totally blows an answer, I won't ask the same question again; the second answer will always be flat. Instead, I'll ask the same question a different way and, generally, a question or two later.

The Interview Structure

Just before the camera starts rolling for a taped interview, I remind the guest to include my question in their answer. More times than not, they do. And, in almost all cases, their parroting of my question is terrible. But, as soon as they repeat the question, they always pause, and, 95 percent of the time, the next words out of their mouth are a perfect start for the answer. I don't know why, it just is. In 40 years of interviewing, I've come to depend upon it.

Sometimes after an almost-but-not-quite-good-enough answer, I'll just ask, "Why?" and stay quiet.

My first question for a recorded interview is always, "Just for the record, but not for the interview, may I have your name, title, and company?" I don't expect to use the answer, but it allows the guest to warm up with an easy answer, allowing them to relax and think to themselves that "This isn't so bad. I can do this."

At this point, the interview dance begins. Time to discover the facts with what, who, and when questions. "What happened?" "Who was involved?" "When did you first notice this?" Avoid any question that can be answered with a yes or no. For example, never use "could," "should," "did," "do," "can," "will," or "have" because, once the guest answers yes or no, you have an answer you can't use, plus you need to scramble to figure out your next question.

The benefit to using what, who, and when questions is that they get into the facts of what happened. They are not asking for opinions. Most of these answers are emotionally neutral, recite the facts, and build common ground. These also set the scene for what's coming next.

After the guest is relaxed, warmed up, and telling stories, it's time to shift the focus to how and why questions. These dig deep into opinion, generally with a lot of emotion attached. If you turn to these too early, the guest will freeze up. Once they realize you are a good listener, paying attention and not trying to trap them, they will flow smoothly into sharing opinions and their feelings about it. You can ask questions such as "How did this make you feel?" "Why was this necessary?" "Why did everything fall apart?"

Sometimes after an almost-but-not-quite-good-enough answer, I'll just ask, "Why?" and stay quiet. Generally, by this time in the interview, the guest is comfortable talking to me and wants to show that they are working with me to create a great interview. That solitary "Why?" causes them to reach down and reveal the emotional core that I'm after for the interview.

At the end of every interview, I always ask, "What question should I ask that I did not?" I don't know everything, and more often than not, this question generates a surprisingly powerful response from the guest. Many times, it becomes the highlight of the interview. As long as your guest feels that they are a partner contributing to this interview, this question can unlock a gold mine of information.

During the interview, I'm always listening for that special sound bite to start an interview and the sound bite that ends it. The start is almost always in answer to a "What...?" question. The close is almost always in answer to "Why?"

At the end of the interview, I always ask: "What should I ask that I did not?"

When the interview is over, the first words out of your mouth after "Cut!" should be to the guest. Even if the interview was a train wreck, congratulate them on doing a great job. Let them know how good they looked on camera. Reassure them that they did a good job.

While there is a fine line between flattery and downright lying, telling a guest they were terrible won't improve your interview. So, you might as well make them feel good as they leave the set. Only after the guest is gone should you discuss technical issues with the crew.

The exception to this is if there is an audio or video problem that occurs during an interview. In that case, I instruct the crew to stop the interview at the end of a guest's answer so that we can fix the problem. Don't interrupt the guest during an answer, because it will destroy their concentration and make them all tense again. They will feel the problem is their fault.

I said at the beginning of this chapter that an interview is a formalized process of sharing information with an audience within a limited time. When you do your homework, set a goal, write down questions, make your guest feel comfortable, and ask questions that generate solid answers, both you and your viewers will benefit. And your guests will think you are a genius!

Don't Be Afraid to Ask Hard Questions

In all interviews, except for people hopelessly out of their depth, part of your role is to challenge their statements, make them think, and ask hard questions. You don't need to be antagonistic; rather, you are representing the audience: What do they want to know? What questions do they want you to ask? Serving "softball" questions benefits no one; push back! Any guest who's done even a few interviews has the ability to explain why the company chose to do something, how this problem started, or what they plan to do to fix it. Your role, as interviewer, is to represent the audience.

Ten Simple Rules to be a Good Interview Guest

My interview career has spanned exploring exotic vacation spots to discussing technology with CEOs. I've lost count of how many interviews I've produced, directed, or hosted; it's well into the multiple thousands, **FIGURE 11.5**.

FIGURE 11.5 As a guest, provide complete answers and follow the lead of the host. (Chris Bobotis is the guest.) (Image Credit: Digital Production Buzz / Thalo LLC)

While I've interviewed my share of celebrities and "professional guests," most of my subjects were amateurs, mostly industry executives who were invited to be guests based on their role in the industry or their company. Many guests were fun and interesting. Some guests could not complete a simple English sentence, and others were sweating so much from sheer terror that I felt sorry simply asking their name.

A few years ago, after completing 104 interviews in four days for a live podcast at a trade show, I wrote a blog called "10 Rules to Be a Successful Interview Guest."[2] While these rules are focused on technology, they actually apply to any business. Remember Dale Carnegie and WII-FM? Your answers should focus on the benefits the *audience* gets from listening to you, not simply how great you are. Here are the highlights of what I wrote.

If you can't explain it in plain English, no one listening will do the work to figure it out.

An interview is a chance for you to explain to the world at large what's important about your company's products and answer questions that a typical user might ask. In other words, you already know everything you need to know for the interview. Here are ten simple rules to make a success of any interview:

- **Who you are.** Be able to explain what your company does in a single, succinct paragraph. You get bonus points if you can avoid using words like "solutions," any acronym with more than three letters, or hyphenated techno-speak. If you can't explain it in English, no one listening will do the work to figure it out.

- **Why you.** Be able to explain why your product, service, or idea is better. Your potential customers are comparing you to your competition, so give them some reasons to consider you. This is also true for any situation where you are trying to improve your visibility. Why should anyone in the audience pay attention to you?

- **Interviews are conversations.** Don't give yes or no answers, even if the interviewer asks yes or no questions. I had one guest who answered

[2] https://larryjordan.com/articles/10-rules-for-a-successful-interview/

every question with "yes" or "no," regardless of how I phrased my questions. On the other hand, I also had three guests who, after I asked my first question, delivered a nine-minute monologue, barely stopping for breath. Neither of these is a conversation. Worse, neither of these is interesting. When the interview was done, I thanked the guest and then had a side conversation with their PR rep to have them rehearse a *lot* more before their next interview. The three who couldn't shut up were never invited back.

Interviews are conversations, not monologues.

- **Have a story.** Come prepared to explain how your product works in the real world. Nothing makes a product more approachable than an example of how it is successfully used. I had a great interview with a CEO of a lens company. We started talking about how lenses were made, and he told a fascinating story of how different varieties of glass can influence how a lens handles light. Very cool. I moved him into a more prominent position in the show based on the interesting content of that one story.

FIGURE 11.6 Stephanie Bauer Marshall is focused on sharing her stories with enthusiasm. (Image Credit: Digital Production Buzz / Thalo LLC)

- **Relax.** Interviews are not life-threatening. No one, I repeat, no one has died during an interview. Go with the flow. If the host is serious, be serious. If the host is cheerful, be cheerful. Enthusiasm is the single best characteristic to bring to an interview. If you can't get excited about your product, no one else will either. Let your personality out a bit. Feel free to be excited, **FIGURE 11.6.**

- **Don't upstage the host.** For me, my guests are the stars. But not all hosts are like me. Take time to figure out what the host wants: stories, high-level strategic positioning, nuts-and-bolts tech, a chance to show off…. Your goal is to get invited back—more visibility is a good thing—not to prove that you are funnier/smarter/weirder than the host.

- **Know your product.** I had four different guests during the trade show this week that had been with their company for only a month or so. They didn't know their products. They didn't know their prices. They didn't know key features. And they won't be invited back. Bring notes, but don't read from them. While this is a conversation, not a lecture, as a guest you still need to know your products or service.

People who listen to interviews are after solid, reliable, practical information they can use.

- **Talk simply to your audience.** In fact, just speak English. You don't need to dumb it down, but if you can't explain it without using a sentence containing more than two acronyms, then practice explaining it until you can.

- **Don't sneak in a commercial.** Any competent host will give you time to promote yourself. Don't try to weasel in a commercial. It breaks the flow and sounds unnecessarily self-promoting. Also, don't do a litany of websites, Twitter, Facebook, LinkedIn, email address, post office box, phone number, booth number, and physical address. Drive people to your website or social media. Keep it simple.

- **Focus on your message.** The host has a goal: to create an engaging interview filled with interesting information. You need a goal, too. And it isn't "buy my products." No one buys a product based on an interview. Instead, your interview goal needs to be, "Make my products so interesting that people have to learn more at my website."

Your role as a guest is to give the host and audience an interesting, compelling interview, filled with intriguing facts and presented by someone who is enthusiastic. I guarantee that when you do, listeners will beat a path to your website to learn more.

KEY POINTS

Here are the key points on creating content through interviews:

- Interviews are a great way to capture and share information with an audience.
- Interviews require planning.
- The host is in charge of the interview.
- Never ask yes or no questions.
- Guests have responsibilities in an interview, as well.

PRACTICE PERSUASION

Interview a friend about their hobby. You don't need to record this. Try different questions and see which ones provide the best answers. Ask them to restate your question in their answer. How does that change the timing and content of their answer? What's the difference in emotional value between who, what, when, where, how, and why questions?

DECODING INTERVIEW-SPEAK

Larry Jordan

Ever wonder what the marketing words mean that guests say during an interview? Here my guide for how to translate the "secret interview" code.

"We are industry-leading, best-in-class" means nothing, while making the CEO feel great. Any "class" can be defined to contain exactly one member; so can "industry" for that matter. When was the last time any company said they sold second-rate goods?

"We have a great set of solutions" means that they read a marketing book somewhere that says all products should solve a problem. The trap with this approach is that it presumes they first *understand* your problem. This is just marketing trying to sound important.

"We provide a one-stop-shop" means that they do a whole lot of unrelated stuff, none of it particularly well, so that you won't discover the competition.

"This is designed for the media and entertainment space" means that it uses either pictures or sound, so therefore *someone* in Hollywood might be interested in this.

"Our product is enterprise-class" means the starting price is $20,000.

"We have great customer support" means that the product doesn't work, so they are nice when you call to get it fixed.

"We are customer-facing" means, um, what? That they have customers?

"We were selected for use in the Olympics/Super Bowl/World Series" means that the starting price is $50,000.

"We are designed for the professional user" means we have an interface that only someone who is compelled to use it for a living will ever figure out.

"We are used in all the major studios and broadcast networks" means that the starting price is $100,000 and requires an army of highly trained IT professionals to operate.

"We have a cool product that we think you'll like" means they have a cool product that's worth checking out.

Good soundtracks aren't just a matter of art. You also have to understand the science.[1]

JAY ROSE
CLIO-WINNING SOUND DESIGNER

CHAPTER 12

SOUND IMPROVES THE PICTURE

In this chapter, we'll define key audio terms, then look at a complete audio workflow from picking gear to outputting audio. The process is the same whether you are creating audio for a podcast, YouTube video, or broadcast, though the specs and attention to detail will be different. Specifically, we'll cover:

- Key creative and technical audio terms
- Selecting the right audio gear
- Recording audio
- Editing audio
- Mixing audio
- Output and compressing audio for distribution

THE BEST WAY TO IMPROVE the quality of your picture is to improve the quality of your sound. Why? Because while the image shows us "reality," audio sparks the imagination and takes us places that video can't go. Even if you are doing an audio-only podcast, the quality of your audio will determine whether your audience thinks it's worth the effort to listen.

The goal: Capture the best audio you can during production, preserve that quality during editing and mixing, then retain that audio quality as you export and compress your file for final distribution.

Another huge benefit to audio is that it drives emotions in a way that images just can't. Images hook the eye; sounds hook our emotions. It's the difference between looking at an image and listening to music. Only one of these gets us tapping our feet and dancing.

Persuasion is more than a message. It is getting the attention of an audience, holding it long enough for them to hear and see your message, then deciding to act on it. *Nothing* turns people off faster than bad audio.

People will happily watch bad video—YouTube is the ultimate proof of this—but they will *never* listen to bad audio. You know this for yourself. How long will you watch a YouTube video with hollow, soft, or distorted sound?

I was reminded of this recently when I went to watch an IMAX film. The image filled an 80-foot screen and looked glorious! But the audio was too loud and distorted. I left after two minutes and demanded a refund.

So, how do we prevent bad audio? Or, more importantly, how do we create good audio? The goal is easy to state: Capture the best audio you can during production, preserve that quality during editing and mixing, then retain that quality as you export and compress your file for final distribution. What could possibly go wrong?

Well, a lot, actually. But, whether you are creating podcasts, motion graphics, or videos, the process is the same.

Glossary of Audio Terms

To get started, let's define some creative and technical audio terms.

Creative Audio Terms

These terms are used to describe the different audio components in an edit, from a creative point of view:

- **Dialogue.** The sound of people talking or narration. Dialogue provides the "facts" and explains what's happening.

- **Sound effects.** Sounds that reinforce what's happening on-screen or in dialogue such as door slams, footsteps, or whooshes. Sound effects are used to make images more believable.

- **Music.** The emotional driver of any scene. Music doesn't need to be a recognizable tune; simple chords are emotionally evocative. Dialogue tells us what to think; music tells us what to feel.

- **Sound design.** The process of creating an aural "environment" of sounds and effects to establish the sound of a world; for example, the "sound" of the starship *Enterprise*. On-screen environments are never silent.

The Power of Audio

If you want to understand the power of audio to persuade, jump to the end of this chapter and read the story about Stan Freberg. Modern media advertising stands upon the foundations created by Stan Freberg and David Ogilvy.

I mentioned that music hooks into our emotions. This is not to say that dialogue or sound effects don't create emotions, because they do. But, within the world of sound, each of these elements plays a different role, all contributing to the meaning and emotional impact of a scene.

Images hook our eye; sounds hook our emotions.

Audio Workflow Terms

These are the four main stages in an audio workflow:

- **Record** (reh CORD). The process of capturing an audio performance.

- **Edit.** Similar to editing text, the process of keeping the good stuff and getting rid of the stuff you don't need.

- **Mix.** Just as different instruments in a band combine to create the finished piece, the process of combining dialogue, sound effects, and music to form a finished work.

- **Export.** The process of converting a mix into a single audio file for distribution and playback.

Technical Audio Terms

There's also a host of technical terms for audio. These are the most important:

- **Sound waves.** How sound travels through the air. Sound waves vary in speed (frequency) and height (volume).

FIGURE 12.1 This is what a waveform looks like. The volume of a sound displayed over time, where peaks indicate louder sounds.

- **Waveform.** A visual representation of the volume of a sound, displayed over time, as illustrated in **FIGURE 12.1**.

- **Volume.** How loud or soft a sound is. Volume is measured in decibels (dB) and displayed on audio meters.

- **Audio levels.** A control that allows you to adjust the volume of a sound, also measured in dB.

- **Frequency response.** The range of audio frequencies from deep bass to high treble, measured in Hertz (Hz) (**FIGURE 12.2**). Not all frequencies can be heard by people. Audio frequencies are logarithmic, which means that each time the frequencies double, the pitch goes up an octave (for you music majors out there). A low E on a bass guitar is 40 Hz. A dog whistle is around 30,000 Hz.

- **Human hearing.** A subset of all audio frequencies that represent the hearing of an 18-year-old person with "normal" hearing. This is described as 20 Hz to 20,000 Hz. Children and dogs can hear above this range. As we age, we lose the ability to hear higher-frequency sounds.

- **Human speech.** A subset of human hearing, ranging from roughly 150 Hz to 8,000 Hz. This also varies by gender and age. This range of frequencies in human speech is important. Vowels—"a," "e," "i," "o," and "u"—are all low-frequency sounds. Consonants are higher-frequency sounds. While human hearing spans ten octaves, human speech covers only about five octaves. In general, women's voices are one octave higher than men's.

In general, women's voices are one octave higher than men's.

FIGURE 12.2 This graphs the frequency range of normal human hearing, from deep bass on the left to high treble on the right. Frequency is logarithmic. (Image © 1985-2020, SOS Publications Group and/or its licensors)

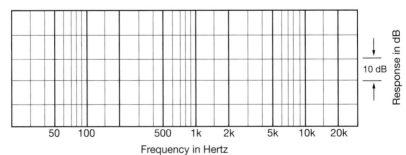

- **Audio channels.** The number of separate audio "tracks" within an audio clip. A monophonic (mono) clip has one channel. A stereophonic (stereo) clip has two (**FIGURE 12.3**).

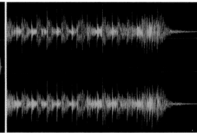

FIGURE 12.3 A mono audio clip (left) and a stereo audio clip (right).

- **Codecs.** Just as we saw with digital images, codecs allow us to convert sound waves from the real world into binary digits that can be stored and played back on the computer. Typical audio codecs include WAV, AIF, MP3, and AAC.

As you'll discover, working with audio principally involves codecs, frequencies, and volume in lots of different and creative ways.

Picking the Right Gear

There are two really important points to remember from this chapter. This is the first: The microphone (mic) you use for audio recording has a greater effect on audio quality than any combination of amplifiers, mixers, recorders, or post-production effects.

Mic technology changes slowly because how our ears work doesn't change. Highly popular, high-quality mics are being manufactured today whose designs haven't changed in 20 years. It is not unusual to go into a recording studio and work with 40-year-old mics. Why? Because they sound great. The

The microphone you use for audio recording has a greater effect on audio quality than any combination of amplifiers, mixers, recorders, or post-production effects.

changes in audio technology are not in how mics work but in how they connect and how they are processed by a computer. That part of the industry is evolving quickly.

The mic must be close to the speaker to get good-quality sound.

In this chapter, I'll recommend specific gear, and the companies that make it, that has proven itself over time. Still, when it comes to audio, nothing beats your ears. If it sounds good to you, it is good.

A key point about all mics is that they sound better the closer they get to your mouth. The best place to put a mic is 1 to 4 inches away from the speaker's mouth. (See the "One More Note About Mic Placement" sidebar in this chapter.) This captures the nuances in the speaker's voice, while minimizing room noise. However, a mic that close will, all too often, spoil the look.

Radio DJs like to "swallow" the mic—get so close to it that it looks like they are eating it. This is because many hand and desk mics have a "proximity" effect. The bass in a voice gets boosted when the mouth is *very* close to the mic; extra bass is considered sexy. A better way to deal with this is to add some bass to the mix after the recording is complete and back away from the mic to avoid pops and clicks.

Microphone Types

There are five categories of mics to choose from, each with their strengths and weaknesses:

- Camera
- Lavalier, or lav
- Headset
- Hand
- Shotgun

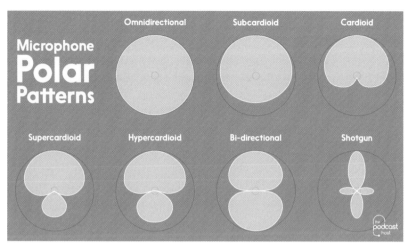

FIGURE 12.4 These are "pick-up patterns," the area from which a microphone will pick up sounds. Notice how shotgun mics reject noise from the side. (Image Credit: ThePodcastHost. com.)[2]

Each of these five mic categories is designed for specific situations, which I'll cover in a minute. As well, each mic has a different "pickup pattern," shown in **FIGURE 12.4**, which is the area where they are most sensitive to sound.

For example, shotgun mics only "hear" sounds directly in front of them. Camera and lav mics, on the other hand, have wide omnidirectional pickup patterns, meaning they hear both voices and room noise equally. This can make for noisy recordings or feedback during a live event. For this reason, lavs are best used in studio or other quiet environments.

One reason why shotgun mics are popular on movie sets, which tend to be noisy, is that they ignore (reject) sounds from the side of the mic. However, this audio "focus" means the audio recordist needs to carefully point the mic at the person speaking for the mic to pick up their voice.

In general, mics with cardioid or super-cardioid patterns are a better option than omnidirectional mics because they tend to reduce noise coming from behind the mic or around the sides.

The camera mic is the easiest to use, but not a good choice. It's easy because it's on your camera. It's bad, because it is located so far from your talent that it picks up more room noise than voice; this makes voices sound "thin" and remote. Camera mics are good for picking up the sounds of the environment, but not people speaking.

Assuming the talent is looking mostly straight ahead, the best place to clip a lav mic is around the second button on a shirt, just below the chin.

FIGURE 12.5 A Sennheiser MKE-2 lav mic. Lav mics clip to a shirt or blouse and are designed to be small and visible. Lavs are heavily used in video. (Image Credit: Adorama Camera, Inc.)

Lavalier mics (**FIGURE 12.5**) are used whenever seeing a mic is OK such as in an interview, newscast, or talk show. They are small, have good quality, and, with proper handling, will last for years. Any lav will sound better than any camera mic. However, feedback is often a problem when used for live events

[2] ThePodcastHost.com: "Learn everything you need to know to start your own podcast." https://www.thepodcasthost.com/equipment/microphone-polar-patterns/

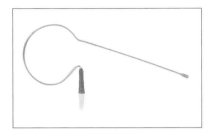

FIGURE 12.6 A Countryman E6 headset mic. (Image Credit: Countryman Associates, Inc.)

FIGURE 12.7 An Electro-Voice RE-20 desk mic shown attached to a mic stand.

FIGURE 12.8 The hand mic is a staple of live performance. This is my favorite: the Shure SM-58. (Copyright © 2009-2020 Shure Incorporated. Image Credit for Figure 12.7: Bosch Security & Safety Systems)

and the speaker's voice is soft. As well, lavs often pick up clothing rustle and room noise.

Headset mics (**FIGURE 12.6**) are ideal for public speakers, theater performances, and other venues where there's some distance between the speaker and the audience. They are also popular for audio podcasts. Headset mics sit about 2 inches from the speaker's mouth, an ideal place to pick up clean audio. But, that also means they are visible. Headset mics are also good at preventing feedback. Countryman, DPA and AKG are all good brands.

Desk mics (**FIGURE 12.7**) are the workhorse of the mic category. From podcasts and narration to radio broadcasts and late-night talk shows, each relies on some form of a desk mic. These provide the richest bass, protection against pops and clicks, and widest frequency response.

Desk mics, meaning mics attached to a stand, are also the dominant mic in the music industry, and there are hundreds of variations to choose from.

Just as picking a favorite color is very much a personal choice, picking a favorite mic is a deeply personal choice; I am a big fan of the Electro-Voice RE-20.

How to Pick the Right Mic for You

When I was setting up my video studio, I was trying to decide which mics to use. So, I rented ten different lavalier mics, gathered a group of friends, and did a blind test to see which ones they liked better. After switching between a variety of mics, the group liked Tram TR-50s for male voices and Sennheiser MKE-2s for female voices.

The moral of this story is that there's no one mic that's perfect for everyone. If you plan to do a lot of recording, do some testing to find the mic that makes your voice sound the best. That's where renting mics for testing can help.

Hand mics (**FIGURE 12.8**) are seen wherever fine performances are presented. Held close to the mouth for best pickup, virtually indestructible to cope with enthusiastic performers, and available in a variety of colors and finishes, these are the Swiss Army knives of audio. Like desk mics, there are dozens of manufacturers to choose from.

At one point, I owned ten Shure SM-58s. A "tight" mic, meaning you need to hold it close to your mouth, the Shure SM-58 is great for recording interviews in noisy environments, surviving an incredible amount of abuse, and having a full, rich sound that really enhances a voice.

The shotgun mic (**FIGURE 12.9**) is the best choice when you want to record audio, but not see a mic, in other words, a film set. The key point to remember about shotguns is that you still need to get them close to the talent—2 to 4 feet from their mouth—but to do so in such a way that the

FIGURE 12.9 A Røde NTG-2 8-inch shotgun mic. These come in two lengths: short (8 inch) and long (18 inch). These are used when you don't want to see the mic. (Image Copyright ©1997 - 2020 B&H Foto and Electronics Corp)

mic is *just* outside the edges of the frame. Most of the time, a lavalier mic will be fine because you don't mind seeing the mic. However, when seeing a mic would look out of place, shotguns are the mics to choose. Shotgun mics come in two lengths: 8 inch and 18 inch. The difference between the two lengths is that the longer mic has a narrower pickup pattern. Reputable shotgun manufacturers include Røde, Azden, Audio-Technica, and Sennheiser.

In situations where it's not possible to get a shotgun mic close enough to the talent to pick up their voice, the audio recordist will either hide a lavalier under the talent's clothing or give up and record the audio later in a recording studio, using a process called "automatic dialogue replacement" (ADR).

Why Pay More for a Mic?

The short answer is that you get what you pay for. More expensive mics have better bass, smoother frequency response, more rugged construction, and better sensors and electronics. Generally, if I plan to use a mic a lot and I'm not hiring myself out as a recording studio, I'll spend $250 to $600 (US) for my main mic. The quality will make me sound better. Recording studios typically spend thousands of dollars for a high-grade mic, such as a Neumann U 87.

Picking the Right Cable

Up until recently, professional-grade mics connected via shielded XLR connectors, while audio gear connected via either XLR or 1/4-inch shielded plugs and cables. (All three types are illustrated in **FIGURE 12.10**.) These were designed for long cable runs, even across power cables, without picking up hum or noise due to their shielding. (RCA connectors are inexpensive, but they are noisy and pick up hum easily because they aren't shielded. Avoid using them if you can.)

FIGURE 12.10 The three main analog audio connectors: RCA (left), 1/4-inch shielded plugs (bottom), and XLR (right). (Image Credits: Wikimedia Foundation (right), Amphenol Tuchel Industrial GmbH (bottom), Wikimedia Foundation (left))

Recently, studios are rewiring using audio over IP (AoIP) signals, which converts the analog signal from a mic or other audio source into Ethernet packets. This provides much greater flexibility in signal routing and, if done right, with no loss of quality. AoIP also delivers far less noise.

Analog-to-Digital Converters

I prefer using traditional XLR mics because this format provides access to the greatest variety of gear, highest sound quality, and reasonable prices. However, these mics are all analog. This means the analog signals need to be converted from the mic into digital signals for the computer. There are two ways we can do this: firmware integrated directly into the mic or stand-alone audio converters.

There are a variety of new mics with analog-to-USB converters built into them so that you connect the mic directly into your computer via USB. Most of these, so far, have been somewhat low-quality devices, though that is changing as the market for high-quality USB mics—especially for podcasting—continues to grow.

My preference is using a separate device called an "analog-to-digital converter." This takes the analog signal from the mic and converts it to USB. (USB 3.x is more than fast enough for transferring audio files.)

There are several of these analog-to-digital converters to choose from. The prices range from $100 to $200 (US). While not free, these units provide extremely high-quality audio at reasonable prices. Brands I've owned and like include:

- Focusrite, shown in **FIGURE 12.11**
- Steinberg
- Behringer
- PreSonus

FIGURE 12.11 An analog-to-digital converter: Focusrite Scarlet 2i2. (Image © 2018 Focusrite Audio Engineering Plc)

I tend to recommend against converters costing much less than $100, as the software used to digitize the signal tends to sound tinny.

Down the Rabbit Hole

Once audio pros start talking about audio, they wax rhapsodic about the quality of audio cables, connectors, and whether gold leaf on a connector really makes a difference. The short answer is that for our work, any reputable cable and connector will be fine; again, avoid RCA connectors because they let too much noise leak in.

Before we leave mics, there is an emerging category of microphones: mics for mobile devices. These include mics from:

- Røde
- Blue
- Shure, shown in **FIGURE 12.12**
- Apogee

These are, most often, lavalier mics. Their biggest benefit is that they connect directly into your smartphone, either via USB or Lightning. Anything that gets the mic close to your talent is a good thing. These mics range between $100 to $200 (US) in price, so, again, if recording audio is something you plan to do regularly and you are using a smartphone, these mics can simplify the gear you are carrying around. As always, evaluate the sound of a mic before committing to it for an important project.

FIGURE 12.12 Mics designed for smartphones are small and connect via USB or Lightning. (Image © 1996-2020, Amazon.com, Inc. or its affiliates)

These mics work the same as the mics we discussed earlier; they simply include the analog-to-digital converter inside the mic, or a belt-pack.

Mixers and Multichannel Recorders

Most of the time, the mic connects directly into the camera, computer, or recorder. However, if you need to record multiple speakers at one time, you have two options:

- Gather all the different mics into a mixer and create a live mix in real time. This is the system used by all live events and performances, but this requires more gear.
- Record every mic to its own channel, using a multichannel digital audio recorder, then sort it out during the audio edit and mix. This is the system used by virtually every movie; however, it requires time spent editing and mixing.

Both are high-quality, both work great; it's just a question of time and budget. If you need the mix done right now, then a mixer is the only option. However, the gear to make your live mix sound good will cost more money than editing. If you have time to edit and mix, multichannel recording is a better choice because it gives you more control over each audio signal during editing and mixing, without spending money for the hardware needed for a live mix.

FIGURE 12.13 The RØDEcaster Integrated Podcast Production Studio is designed for podcasters and smaller (up to four mics) productions. It has high quality and lots of flexibility. (Image Copyright ©1997 - 2020 B&H Foto and Electronics Corp)

I used a Mackie 1402VLZ mixer for my live podcasts for almost ten years. Then, I switched to a Behringer (with significant bumps during installation and operation). Podcasting friends swear by the Røde mixer, pictured in **FIGURE 12.13**. All these mixers are designed for live streams, podcasts, and broadcasts.

You still need speakers and probably a pair of headphones. But you know how to use your ears and find ones that suit your budget and taste.

Digital Audio Recording

Now that we have all our gear connected and running, we need to record it. When it comes to recording, we have three options:

- Record the audio to the camera
- Record the audio directly on the computer
- Record the audio on a separate audio recorder

Recording audio to your camera is the easiest, but it limits you to only two mics. Also, camera audio circuits are not always the best. This is a good option in a pinch, but not preferred.

A Big Caution!

When buying headphones or monitor speakers, it is important to buy ones that have a flat frequency response. These use the word "monitor" somewhere in their description. You may love listening to audio with a thumping bass track, but when you mix, you want to hear what's actually in your mix, not what the speakers are adding. Your listeners won't have the same speakers as you do.

Listen to audio on your personal time using whatever headphones or speakers you want, but *mix* audio using monitor speakers that provide a flat frequency response. Reputable audio manufacturers include a chart showing the frequency response of their speakers so you know what you are buying. Monitor speakers that I like include Yamaha, M-Audio, and JBL. If you have the money, Genelec speakers are outstanding.

Recording audio directly to your computer is the best choice for quality, especially if you are doing an audio-only podcast. It's how I recorded hundreds of interviews—directly to my computer using either mics or Skype. But, this also requires having a computer on set to record with.

Recording to a separate digital audio device provides the highest amount of portability combined with excellent quality. I'm a fan of Zoom, Marantz, and Sound Devices gear. These span the range from not too expensive to high-end pro gear. My personal go-to audio recorder is a Zoom H4n (**FIGURE 12.14**).

Any stand-alone audio recorder should provide support for multiple microphones, separate gain control for each mic, and record to an uncompressed audio format, such as WAV or AIF.

Most of the time, when you record audio, you are able to set recording levels. Because we don't want our audio to "distort," which sounds scratchy, crackly, and terrible, I recommend setting audio recording levels such that the loudest portion of the speaker's voice does not exceed –12 dB. (Yes, audio levels are measured in negative numbers!)

Set audio recording levels such that the loudest portion of the speaker's voice does not exceed –12 dB.

FIGURE 12.14 The Zoom H4n Pro, a stand-alone digital audio recorder. (Image © 2020 Zoom North America)

This allows the audio to be recorded at a good volume, while protecting against distortion in case someone starts shouting. Depending upon your gear, you are often able to monitor levels during the recording, which is something I recommend you do.

The closer the mic, generally, the better the bass and less the noise.

Also, just as reassurance, the level we record audio is never the same as the final mix. The final mix is always louder. Why? Because it is easier to increase audio levels during the mix than to remove distortion because our audio levels were too loud during recording.

Here are some good rules for recording high-quality audio on set:

- Always get the mic as close to the talent as possible.
- The closer the mic, generally, the better the bass and less the noise.
- Never use the camera mic for anything other than environmental sounds.
- Always do a test recording to check levels and audio quality (make sure you don't have any hum or strange noise).
- Set the recording level to peak around –12 dB.
- If the unit has an automatic gain control (AGC), you'll most likely get a better recording by turning this off.
- Use desk mics for audio-only recording and lavaliers for video recording, unless you don't want to see the mic in which case use a shotgun.

One More Note About Mic Placement

Never put the mic directly in front of a speaker's mouth. As an experiment, put your hand directly in front of your mouth. Then say "p," "t," "k." Feel the puffs of breath hitting your hand? If you replaced your hand with a mic, those puffs of air would sound like explosions, which is why we call those letters "plosives." ("d", "g," and "b" are similarly explosive.)

Now, move you hand to the side of your mouth and say those letters again. This time, no puffs of air hit your hand.

Note how the host is holding her mic to the side for this interview. This avoids plosives and other audio problems.

What's really helpful about audio is that it radiates in a dome from your mouth into space. Placing a mic to

the side (or above or below) your mouth—as shown here—captures the same quality sound as if the mic were directly in front, without worrying about plosives ruining your audio.

Pop Filters Keep Audio Clean

A pop filter, illustrated in **FIGURE 12.15**, is a thin piece of porous fabric placed between the speaker's mouth and the mic to block plosives and other pops. Sounds pass through; puffs of air don't.

So, you ask, if that's true, why do singers on stage sing directly into a vocal mic? First, I think it's the security of knowing exactly where the mic is and having something to hang onto while performing. Second, many vocal mics include internal pop filters to minimize plosives. Third, audio quality for live concerts is not as high as audio recorded in a recording studio. For proof, look at how the mics are configured the next time you watch a behind-the-scenes video on recording narration or music in a studio. Studio mics all use pop filters between the singer and the mic, as well as placing the mic about 6 inches in front of the mouth and slightly above, specifically to prevent strong plosives from interfering with the recording.

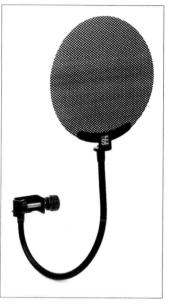

FIGURE 12.15 A pop filter for a mic. (Image © 2020 Sweetwater)

Editing Audio

Whether you record a podcast, a vocal performance, a testimonial, or a drama, editing will always be the next step. We need to get rid of the parts we don't like and keep the ones we do. While it is way beyond the scope of this book to explain all the details of audio editing, just as our chapter on editing video will be equally succinct, here are the highlights.

Remember our definition of audio channels earlier? Here's where these come into play, along with a few more technical terms that apply to editing and mixing:

- **Audio channels.** We defined this earlier as the number of separate audio tracks in a clip. Essentially, each mic records to its own channel. So, some audio clips may just have a single voice in a mono clip, or music in a stereo clip, or still other clips might have 16 voices, each recorded to a separate channel and stored in one clip!

- **Sample rate.** The sample rate is the number of digital bits recorded each second for any given audio signal. Those bits are created when an analog signal is converted to digital. Audio recorded for video has standardized on 48,000 samples per second. (We commonly call this "48K.") Sample rate determines frequency response.

Sample rate also determines frequency response. The Nyquist theorem states that dividing the sample rate by 2 equals the maximum high frequency response of an audio clip. And, no, that won't be on the quiz. Well, probably not. Perhaps.

- **Bit-depth.** (Don't confuse bit-depth with sample rate.) Bit-depth is a measure of the amount of variation between the softest and loudest potential passages in a clip. This is most relevant in orchestral work, which has a far wider dynamic range than, say, rock. The greater the bit-depth, the greater the variation between the softest and loudest passages. Audio recorded for video has standardized on 16-bit depth for recording and 32-bit depth for editing and mixing.

- **Trimming.** This means adjusting the position—longer or shorter—of where a clip starts or ends.

- **In-point.** This is the location where an edited clip begins, which may or may not be at the start of a clip. We almost always refer to this as the "In."

- **Out-point.** This is the location where an edited clip ends, which may or may not be at the end of a clip. We almost always refer to this as the "Out."

Bit-Depth in Audio

Bit-depth in audio, similar to bit-depth in images, determines the dynamic range of a clip, the difference between the softest and loudest passages. The downside is that as bit-depth increases, file size also increases. However, audio files are small compared to video files, so increasing the audio bit depth doesn't really matter a whole lot when considering the overall storage needs of your projects.

TABLE 12.1 Audio File Sizes to Store One Hour of 48K WAV Audio[3]

	MONO	STEREO
8-bit audio	164.8 MB	329.6 MB
16-bit audio	329.6 MB	659.2 MB
24-bit audio	494.4 MB	968.7 MB
32-bit audio	659.2 MB	1.287 GB

All audio screen shots are from Adobe Audition 2020.

FIGURE 12.16 shows a typical, two-person, two-channel audio recording. The host is on the top track, which is Track 1, and the guest is on Track 2. The two channels are in sync, which means they are locked together in time, but each channel can, and most often does, have different audio levels, trimming, and filters.

[3] Source: https:// www.colincrawley.com/ audio-file-size-calculator/

Figure 12.16 illustrates that tall (or thick) wave-forms indicate where a person is talking, and short (or thin) waveforms indicate the quieter sections where that person is not talking.

As we play the audio, the audio meters (**FIGURE 12.17**) show us the absolute level of the sound. 0 dB is clearly marked, with room above to indicate excess audio levels and room below to indicate how levels change during playback.

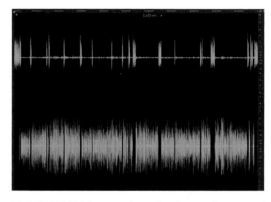

FIGURE 12.16 This is a two-channel audio recording prepped for an edit; the host is on the top track, while the guest is on the second track.

Editing audio consists of three basic steps:

1. Add all the audio needed for the mix into the timeline.
2. Rearrange clips to tell the best story.
3. Trim the edges of clips to create smooth transitions, eliminating audio pops or verbal stutters.

In general, whether editing audio or video, concentrate on telling your story first and worry about trimming and levels after you've got the story organized the way you want. Why? Because it saves time. Trimming takes time to make edits sound right. If you start trimming too soon, then discover that you don't really need that clip, you just wasted time that would have been better spent storytelling.

When it comes to telling a story, it is really helpful to know what you have to work with. In the old days, back when written transcripts were expensive and took a couple of weeks to deliver, editors got in the habit of listening to their clips, taking extensive notes, and hoping they could keep track of everything in their heads.

FIGURE 12.17 Audio meters measure the absolute volume of audio. Note the dB levels marked on the right.

PEAKS The instant-by-instant maximum level of the sound.

Audio Meters Measure Peaks

All audio and video editing software measures audio using decibels relative to full scale (dBFS). While many professional audio mixers prefer measuring levels using average audio levels measured in loudness, K-weighted, relative to full scale (LKFS), most audio and video editing software displays levels as "peaks." Peaks are the instant-by-instant maximum level of the sound. There's about a 20 dB difference between peak and average levels, which makes for some heated conversations if you don't clarify which levels you are talking about ahead of time. In this book, I'll be referring to peaks.

Today, with cloud-based online transcription services popping up like daisies, creating a rough transcript of an audio interview is a matter of just a few minutes and a dollar or two. Among others, popular services include:

- Amazon Transcribe
- Google Speech to Text
- Rev
- Trint
- Speedscriber.com
- SimonSays.ai
- Transcriptive.com

If you aren't working with scripted material, you can speed your edit by creating a written transcript of all your recordings, then do a "paper-edit" where you determine your edits using the text first. Editing text is much faster—and simpler—than working with media.

Once you figure out the story, it's time to edit the clips into a timeline for placement, trimming, and mixing, as shown in **FIGURE 12.18**.

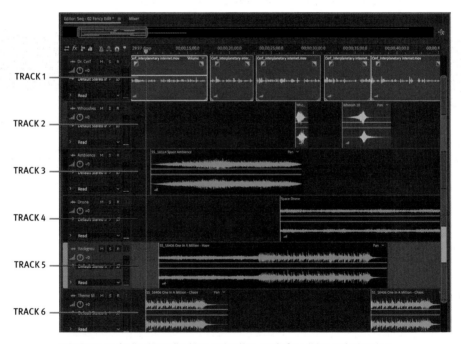

FIGURE 12.18 Audio clips edited into a timeline, ready for editing and trimming.

If you are editing an audio-only program, it will be easiest to edit the clips directly in Adobe Audition or Avid ProTools, which are two popular Mac-based programs. (Personally, I'm a big fan of Audition.)

If you are editing audio that was recorded as part of a video, it is a better idea to edit the audio and video together in a video editing program such as Apple Final Cut Pro X, Adobe Premiere Pro, iMovie, Clips, or Premiere Rush. Then, after the video edit is complete, bring the audio into an audio editor for the final cleanup and mix.

This audio cleanup, which the pros call "sweetening," isn't required, but once you hear the difference, you'll be a fan of using these techniques for your projects.

As you edit, it is important to put similar clips on the same track. For instance, I put male dialogue on Track 1, female dialogue on Track 2, sound effects starting on Track 3, and music starting on the first track after sound effects. The reason for this organization is that levels and effects are generally applied to the *track* that holds the clips, rather than the clips themselves. By grouping similar clips on the same track, we can apply the same effect to all of them, which is more efficient in terms of time and computer processing.

Time spent making your edit sound organized, smooth, and clean pays dividends when the audience hangs around to listen to all of it.

In Figure 12.18, I organized the clips in this same fashion: dialogue on Track 1, effects on Tracks 2 to 4, and music on Tracks 5 and 6. Note that only the dialogue is mono; effects and music are stereo.

Now that the clips are organized and positioned in the timeline, it's time to trim. Trimming simply means dragging the beginning or ending edge of a clip left or right to reveal or hide a piece of audio, as illustrated in **FIGURE 12.19**.

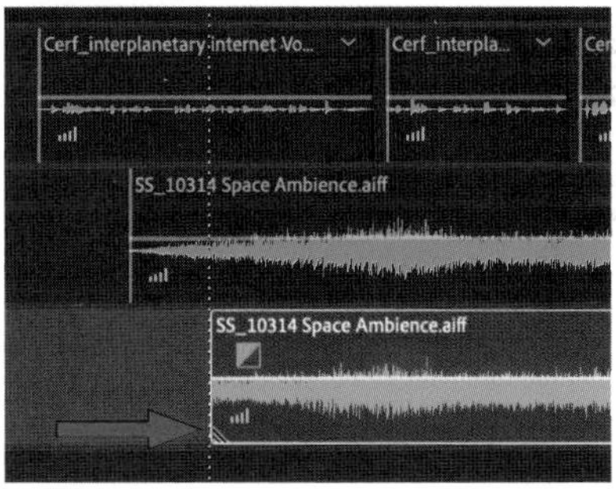

We can also cut clips in the middle, exactly the same as highlighting and deleting text in a word processor, to get rid of a cough or stutter.

FIGURE 12.19 Trim a clip by dragging an edge left or right.

This is an important point. The reason we go through all the effort of editing and trimming is that we want the audience to focus on our story, not the distractions of stutters, clicks, or disjointed sentences. Only by listening to the story, not the mistakes, can we persuade someone to do what we suggest.

Clean audio is more persuasive because people won't turn it off in the middle. Time spent making your overall edit sound organized, smooth, and clean pays dividends when the audience hangs around to listen to all of it.

Mixing Audio

Once the story is complete, recording and editing are done, and trimming got rid of all the junk, it's time to create the final mix. The mix is where sound effects and music are added, final levels are set, and audio effects are applied to create the final sound of our program.

Audio volumes increase as more clips play at the same time.

Mixing, like any creative process, is part art and part technology. The art you learn by doing, while understanding the technology behind it can help improve your art.

Speakers Are Your Friend

When you create your final mix, make a point to listen to your audio on speakers, not headphones. Often, headphones are too good; they provide perfect stereo separation. You will hear things in headphones that can't be heard when using speakers.

Speakers are the preferred mixing tool unless you know that your audience will be listening to your mix on headphones.

Set Audio Levels

The first step in any mix is to set volume levels. We do this by dragging the thin horizontal line in the middle of each clip up or down (**FIGURE 12.20**). As you do, the audio meters will show the results. When creating a mix, we want the audio meters to consistently peak between –3 dB and –6 dB.

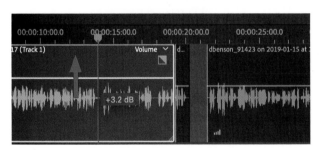

FIGURE 12.20 Drag the horizontal line inside each clip to change audio levels: Up increases volume, and down decreases it.

Unlike working with images in Photoshop, audio levels are additive. This means that the more audio clips that are playing at the same time, the louder the levels will be. In comparison, images in Photoshop don't get brighter if more layers or elements are added to the composite. While the math behind audio levels is complex, it is enough to know that your volumes will increase as more clips play at the same time.

We can also add fades to the edges of a clip by dragging the gray boxes at the top corners of each clip to make the sound ease in gradually or end gradually, as shown in **FIGURE 12.21**. I use fades a lot to minimize strong breaths just before a speaker starts talking or to have the music slowly fade away at the end of a piece. We can also change the shape of the fade to make it ramp levels quickly or slowly, depending upon what we want the audience to hear.

Before you start your mix, you need to decide whether you are creating a mono or stereo file. A mono file takes your audio and plays it equally out both the left and right speakers, which puts the sound in the center, between the two speakers. A stereo file plays Track 1 out the left speaker, and Track 2 out the right speaker, creating the illusion of sonic "space," also called the "sonic field," between the two speakers.

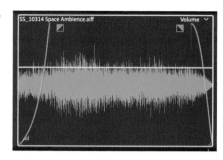

FIGURE 12.21 Add fades to a clip, indicated by yellow lines. Change the shape of the fade to change the speed and sound of the fade.

A common way to mix is to record a variety of mono sources, such as people speaking, room noise (called "room tone"), and on-set sound effects. Then create a stereo mix, so you can pan sound sources around the sonic field. However, not all mixes need to be stereo. Especially for podcasts, stereo is not the best choice.

Let me explain. If you are recording a person talking, say for an interview, podcast, or nonmusical event, how many mouths do they have? Correct. One. Next, how many mics should you use to record one mouth? Correct again. One. Since each mic records to its own channel in the audio clip, if you are editing the audio recorded from one mouth using one mic, how many channels are in your audio editor? Correct for the third time! One.

If you have a dialogue between two people, you aren't going to put one person on the left channel and the other person on the right. What happens if your listener's right speaker is broken? They'd lose all that speaker's audio. Again, you'd pan both speakers voices to the center. This is where they would appear naturally if you were using mono, not stereo audio.

A general rule: Create a stereo mix only if you need to take advantage of the sonic field, that is, the spread of sound between the left and right speakers.

Relative vs. Absolute Levels

When we adjust the audio level of a clip, we are making a "relative" adjustment. We are increasing or decreasing the audio volume *relative* to the level at which it was recorded. We tend to bring sound effects and music levels down, while raising dialogue.

The audio meters show the *absolute* audio level. This is the precise volume associated with playback of the clip or mix at that instant. The audio meters are after—"downstream"—of all audio level adjustments and effects.

During any mix your ears get tired. Always believe the meters; they never lie.

So, the general rule is that you need to create a stereo mix and stereo output only if you need to take advantage of the sonic field, that is, the spread of sound between the left and right speakers. This perfectly describes music, where you place different instruments in different locations, or in drama where you want to emphasize an actor coming in from the left side and moving right.

Your goal, when creating an audio mix, is to have peak levels bounce between –3 dB and –6 dB on the audio meters when all clips are playing.

But, if all you have is one person talking, there's no reason to use stereo. (And, yes, I'll still select mono if I'm using music under their voice, unless the spread of the music across the sonic field is important to the message I'm trying to convey.)

Oh! There's another reason for using mono—for web work, it's the most important. Mono files are half the size of stereo and load twice as fast.

FIGURE 12.22 shows a simple mix. The dialogue at the top tells the story. Sound effects are in the middle, while music cues are at the bottom. Levels are roughed in. Clips are synced and trimmed. All sound effects have fades. It's time to finish the mix.

In the final mix, we apply different effects to create the overall sound we want. While filters and effects are important, nothing is more important than audio levels themselves. As I mentioned earlier, if our audio levels exceed 0 dB at any time during final export, our audio will distort. Not only does this sound terrible, but an audience won't bother to listen to it.

To prevent distortion, my recommendation is to first adjust the levels of all your dialogue, then sound effects, and finally music.

How to Control Pan

You adjust pan (left to right placement of audio) by track, not clip. Pan settings range from full left to full right. Pan does not apply to mono projects; in fact, the control itself does not appear. Here, I'm panning all the clips on this track slightly to the left. (A full pan to one side or the other equals ±100.)

Mono or stereo is determined in the mix. All mics, by default, record mono—one voice into one mic stored in one channel of an audio clip. Yes, there are stereo mics, but each also records only one signal per channel.

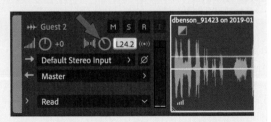

The pan control shifts the sound between the left and right speakers.

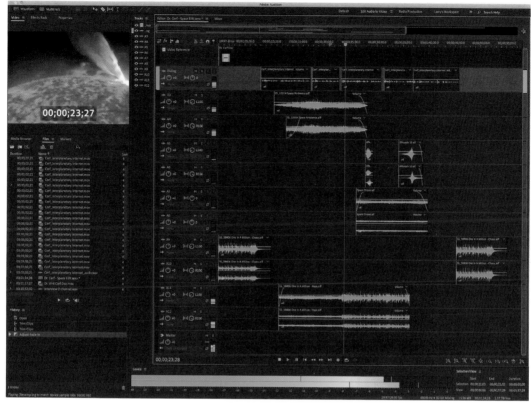

FIGURE 12.22 A simple mix, with dialogue, sound effects, and music that contains eleven audio tracks and one video track. (Image of sun courtesy of Nasa)

Your goal, during the mix, is to have peak levels bounce between –3 dB and –6 dB on the audio meters when all clips are playing. (This equates to an average level of about –16 LKFS, which is a good level for the web.)

Add Audio Effects

After we have the levels roughly set, we can add audio effects. As you can imagine, there's a wide variety of audio effects we can apply. For this chapter, I want to highlight two that will make your dialogue sound a *lot* better.

At the beginning of this chapter, I said that there were two important concepts to remember from this chapter. The first was that the mic you use for recording your audio has more impact on audio quality than any combination of amplifiers, mixers, recorders, or post-production effects.

Here's the second: When you export your final mix, peak audio levels must *never* exceed 0 dB, as measured on the audio meters. Never. Otherwise your audio distorts.

> *When you export your final mix, audio levels must never exceed 0 dB; otherwise your audio distorts.*

During editing and mixing, levels may jump over 0 dB, because we aren't paying attention. As long as you aren't exporting your final mix, you are fine. But, if audio levels exceed 0 dB during final export, your audio will distort. It will sound awful. Listeners won't listen. And it takes a lot of expensive technology to repair it; in fact, up until a couple of years ago, we couldn't repair it at all.

As we saw earlier, we can adjust audio levels by dragging the volume line (the rubber band thingy) in the middle of the clip up or down. This works fine, but it takes forever.

Instead, if you did a reasonably good job recording the audio, it is easier to use a filter to handle the audio levels for you. In Final Cut Pro X, I recommend the Limiter filter. For ProTools, Audition, and Premiere, I recommend the Multiband Compressor, even though the interface is enough to scare small children.

Data Compression vs. Audio Compression

There are two types of compression: data compression and audio compression. Data compression removes data to make files smaller. Audio compression alters audio levels to reduce the variation between the loudest and softest passages. Audio compression does not remove data, nor does it change the size of the file.

What the Multiband Compressor does is raise the volume of softer passages in your audio, without altering the louder passages. It does this on an instant-by-instant basis, without requiring any help from you, aside from initial configuration. Even better, it raises different frequencies differing amounts to make voices sound more robust and "up front." One of the key benefits to using this effect is to prevent any distortion from excessive levels.

In general, you only apply the Multiband Compressor to dialogue, because music has already been aurally compressed and sound effects don't need it. Also, as a general rule, you want dialogue to sound clean and unaffected. So, don't add things like reverb, flanging, choruses, or all the other cool effects lurking in your audio software. Save those for music, cartoons, and other nonstandard audio mixes.

To apply the filter, select the track containing dialogue, then apply the effect. (The precise steps will vary by software—however, effects are always applied to individual tracks or buses.)

Oh. My. Goodness! The first time I saw this plug-in, shown in **FIGURE 12.23**, my thought was: "I am not smart enough to use this!" Don't worry, there are only three settings you need to change:

- Change the Presets menu to Broadcast.
- Change Margin (on the right) to −3 dB.
- Uncheck Brickwall Limiter.

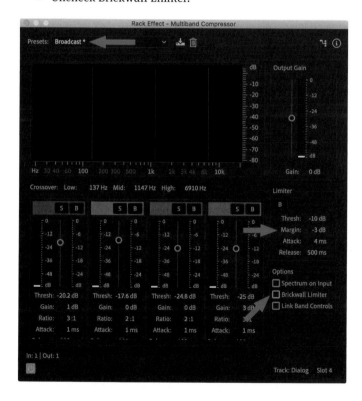

FIGURE 12.23 Don't panic! This is the Multiband Compressor in Adobe Audition, with custom configuration settings applied.

Now play your audio. See how much stronger the dialogue sounds? It's just amazing! Turn the effect off and compare the difference. The really cool thing is that it also evened out the levels in the dialogue without you having to do a lot of personal tweaking.

The Multiband Compressor is probably the most useful filter to learn, which is why I presented it first. But, there is one more filter that can help your mix: Parametric EQ. (EQ is short for "equalization.") This allows us to adjust specific frequencies in a sound. Just as we can shape specific colors in our images to create a specific look, we can "shape" specific frequencies in our audio, as illustrated in **FIGURE 12.24**. Most of the time, we use this shaping capability to create ear-catching sound effects.

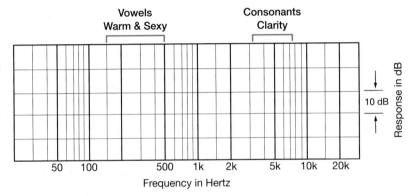

FIGURE 12.24 This chart illustrates both the range of human hearing and typical ranges for vowels, which provide warmth, and consonants, which provide clarity. (Image © 1985–2020, SOS Publications Group and/or its licensors)

Adjusting frequencies has a wide variety of uses, but what we want to do here is "warm up" a voice by boosting the lower frequencies in a voice. Then, to improve clarity, we'll boost a range of higher frequencies. The tool we use to accomplish both these tasks is the Parametric EQ filter.

This frequency manipulation is especially important when creating mixes for older audiences whose hearing may no longer be acute. One of the sad facts of life is that as we get older, we lose the ability to hear high-frequency sounds, which means that it becomes harder to understand what people are saying. Adjusting the EQ can help overcome this.

Effects Are Applied from Top to Bottom

Audio effects are applied in order, from the top of the stack (the list of effects) to the bottom, or, if effects are arrayed horizontally, from left to right. Because of this, the Multiband Compressor must always be last, at the *bottom* of any effects you apply, to protect against distortion. I discussed it first, but, in real life, it is applied last, after all other effects are added.

The Parametric EQ filter will be found with other EQ effects. Apply it the same way you did the Multiband Compressor.

When you display the Parametric EQ filter, it isn't quite as intimidating as the Multiband Compressor, but it's close (**FIGURE 12.25**). Fortunately, there are only a few settings you need to change, and they vary depending upon whether you are modifying male or female voices. (As a general rule, I would not apply the Parametric EQ to children's voices. They are already high and squeaky enough.)

Figure 12.25 illustrates settings for a typical male voice (top) and a typical female voice (bottom). Both settings warm a voice, while adding clarity to consonants.

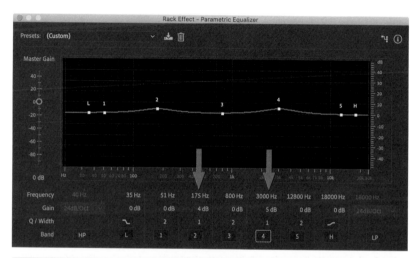

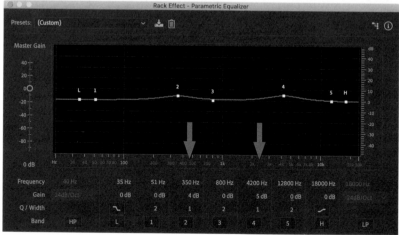

FIGURE 12.25 This is the Parametric EQ filter with settings configured for "normal" voices: male (top) and female (bottom).

In case you are curious, middle C on an 88-key piano has a frequency of 261.6 Hz. Most instruments tune to the A above middle C, which has a frequency of 440 Hz.

TABLE 12.2 illustrates how to adjust settings in the Parametric EQ filter for different voices.

TABLE 12.2 Parametric EQ Settings for Dialogue

SETTING	MEN	WOMEN
Change 200 Hz Frequency to...	175 Hz	350 Hz
Change 200 Hz Gain to	3–5 dB	3–5 dB
Change 200 Hz Q/Width to...	1	1
Change 3200 Hz Frequency to...	3000 Hz	4500 Hz
Change the 3200 Gain to...	4–6 dB	4–6 dB
Change the 3200 Q/Width to...	1	1

Q determines the shape of the curve when applying EQ to clips. The higher the Q, the sharper the curve. Generally, lower Q values (0.8–2) work better for voice EQ.

The Parametric EQ filter exists in Audition, Premiere, and ProTools. In Final Cut Pro X, it's called the Fat EQ filter. The settings and operation are identical.

What these settings do is boost the low-frequency vowels to make voices sound sexier, warmer, and more inviting; then boost high-frequency sounds to make consonants crisper and more audible. In both cases, changing the shape of the curve (Q) smooths these changes by spreading them across a wider range of frequencies.

Working with frequencies is similar to modifying grayscale values; we can't boost a single frequency. We are always adjusting a range—sometimes a wide range, sometimes a narrow range, but *always* a range. Increasing or decreasing the EQ is not a bad thing—it is done all the time. The key, though, is to not do this to excess. It's like cooking; a little spice goes a long way. For instance, adding too much low-frequency EQ makes a voice "rumbly," muffled, and hard to understand.

There is no rule that says you have to use any of these settings. Often, I'll boost the highs to improve clarity and leave the low frequencies alone. As with all things in audio, listen to the results and decide for yourself whether you like it.

Feel free to shift frequencies, adjust the Q, or modify the gain to get the sound you want. My numbers are starting points, not set in stone.

You can always export a compressed version, but never compress an already compressed file. The quality will sound poor. This is why exporting a master file to use for all future compression is a better choice.

Output and Compression

Now that your mix is complete, it's time for final output and compression. (And this time, I mean "data compression" to make the final file smaller.) Before creating your final masterpiece, listen to it one more time to make sure it sounds okay. Verify audio levels don't exceed 0 dB, that all transitions sound smooth, and that your content makes sense. I can't tell you how many times I finished a mix, exported it, then discovered a mistake that I should have caught far earlier.

My recommendation is that you should always export a project as uncompressed audio, meaning either a WAV or AIF file. This provides a high-quality master file that can be archived or used to create compressed versions for distribution on social media.

There are two uncompressed (which means "highest quality") audio codecs: AIF and WAV. Both have the same audio quality, though the industry seems to be migrating toward WAV at the moment. **FIGURE 12.26** illustrates typical settings for a high-quality WAV export.

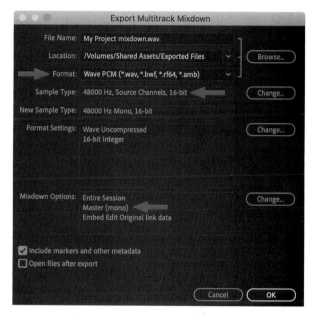

FIGURE 12.26 Export settings for a mono WAV file from Adobe Audition.

Here are the key settings:

- Format: WAV (or AIF)
- Sample rate: 48000 Hz
- Bit-depth: 16-bit
- Pick Mono or Stereo to match your project and mix

Once your master file (which is what I call the final, high-quality export of the audio mix) is exported, you either need to import it into a motion graphics project to serve as your soundtrack, import it into a video as your final audio mix, or compress it for posting to social media. (I'll cover the first two options later in this book.)

Here's an important note about all social media, such as YouTube, Facebook, or Instagram. They *always* recompress your media for compatibility with the different media formats that exist today. I read somewhere that YouTube creates 20 different versions of every video we upload, which is what's happening when you see YouTube "processing" your video.

MASTER FILE The final, high-quality export of the audio mix.

48K or 44.1K?

If you are compressing files for streaming or a download from your own website, there is no sound quality difference between using a 48K or 44.1K sample rate; however, a 44.1K compressed audio file will be 10 percent smaller than a 48K file. I use 48K for video and social media, and I use 44.1K to create a smaller file for my own website. Both 48K and 44.1K exceed the high-frequency response of human hearing.

FIGURE 12.27 Mono audio compression settings for a local website in Apple Compressor.

FIGURE 12.27 illustrates suggested compression settings to help you compress your files, without damaging sound quality. **TABLE 12.3** breaks out compression settings based on what you are going to do with the audio.

You could further tweak these settings to compress these files even smaller, but given the extent of broadband Internet access, there isn't the need to squeeze files as small as we did just a few years ago. (You could also compress your audio files into the AAC codec. The compression values are the same, but given the wide support for MP3, there's no real benefit to doing so.)

TABLE 12.3 Audio Compression Settings by Destination

DESTINATION	CODEC	COMPRESSION SETTINGS
Video editing app	WAV	Don't compress
Social media	MP3	Sample rate: 48 K
		Stereo: 384 Kbps
		Mono: 128 Kbps
Streaming/website	MP3	Sample rate: 44.1 K
		Stereo: 128 Kbps
		Mono: 64 Kbps

KEY POINTS

Whew, this chapter covered a lot! What's important, though, is how this builds on what we've already covered, and it will appear again, in slightly different form, when we get to video editing in a few chapters.

Here are the key concepts in this chapter:

- The microphone you use to record your audio is the most important audio decision you can make.
- Microphones are designed for different voices and tasks. Lavaliers are commonly used for most video work. Use desk mics or headsets for audio-only recording.
- All audio is composed of frequencies and volumes, which vary constantly.
- Codecs are at the heart of audio, just as they are for still images.

- Always do a test recording to make sure everything is working properly.
- Record audio with peaks bouncing around –12 dB.
- Mix audio with peaks bouncing between –3 dB and –6 dB.
- When you export your final mix, peak audio levels must never exceed 0 dB to prevent distortion.
- Edit your audio so it tells a story first; then worry about all the tech.
- Most podcast projects are better mixed to mono, rather than stereo.
- Adjust export and compression settings based upon where you are sending the file.

One last thought. Audio, like video, is endlessly fascinating and very, very deep. Audio sparks the imagination in ways video doesn't. Allow yourself time to play and you'll find limitless ways that audio can make your images look better.

When it comes to persuasion, audio is in a class by itself.

Just ask any podcaster.

PRACTICE PERSUASION

Based on what you learned in the previous chapter about structuring an interview and picking the right questions, use your cell phone and record an audio interview with a friend about their hobbies.

Hold the microphone close to their mouth; then move it 5 feet away and then 10 feet away. How does the quality of the sound change? Which distance provides the greatest sense of "being there"? How do emotional values change as the distance from the subject changes? Does the sound change if you are 3 feet in front of them or to the side or above or below them?

PERSUASION P-O-V

STAN FREBERG EXPLAINS RADIO

The following story was taken from a master's thesis on Stan Freberg, written by Ronda Beaman Martin at Texas Tech University. She had become personal friends with Stan Freberg, who provided the source material for her thesis.[4]

[4] (Martin, 1986)

Stan Freberg (1926-2015) was an American author, actor, recording artist, voice artist, comedian, radio personality, puppeteer, and advertising creative director. (Anonymous publicity photo.)

On May 1, 1985, the U.S. Congress passed a resolution honoring "one of the great satiric and comedy minds of our time, Stan Freberg." The resolution continued: "His humorous and satiric commercials pioneered the way for those advertisers who are now able to sell a product and amuse the public at the same time."

Winner of 21 Clio Awards, the ad industry's equivalent of an Oscar, Freberg developed the humorous, laidback approach to broadcast advertising, giving the listener or viewer credit for more IQ than other ads of the time.

As Freberg said, "If a company does a funny spot, it's obviously not taking itself too seriously, right? It must have a good product or else it couldn't afford to kid around. That's the theory."

Probably his most famous radio ad was to promote advertising on radio. Why? Because he wanted to prove that audio could create images that would take an immense budget for video to equal. When it comes to sparking the imagination, audio is unmatched. Here's his script—remember this is audio only, with *lots* of sound effects (SFX).

Voice #1: Radio, why should I advertise on radio? There's nothing to look at, no pictures.

Voice #2: You can do things on radio that you couldn't begin to do on TV.

Voice #1: That'll be the day.

Voice #2: All right, watch this:

Hm...OK, people, now, when I give you the cue, I want the 700-foot mountain of whipped cream to roll into Lake Michigan, which has been drained and filled with hot chocolate. Then, the Royal Canadian Air Force will fly overhead towing a 10-ton maraschino cherry, which will be dropped into the whipped cream, to the cheering of 25,000 extras.

All right, cue the mountain. (SFX)

Cue the Air Force. (SFX)

Okay, 25,000 cheering extras! (SFX)

Now, you wanna try that on television?

Voice #1: Well...

Voice #2: You see, radio is a very special medium, because it stretches the imagination.

Voice #1: Doesn't television stretch the imagination?

Voice #2: Up to 21 inches, yeah.

If we substitute "audio" for "radio" and even expand the size of the TV monitor, the truth of Stan Freberg still holds.

Stan Freberg introduced satire to advertising and revolutionized radio by influencing staid ad agencies to incorporate his style into their previously dead-serious commercials. (Image courtesy of ABC Television.)

Cinema is a matter of
what's in the frame
and what's out.[1]

MARTIN SCORSESE
FILMMAKER

CHAPTER 13
VIDEO PRODUCTION

We covered lighting and composition in Chapter 7, "Persuasive Photos." We discussed audio in Chapter 12, "Sound Improves the Picture." This chapter begins where Chapter 10, "Video Pre-production," ends: Planning is done, we now need to convert our plans into pictures. More specifically, this chapter looks at video gear and talent staging. The goals of this chapter are to:

- Review production planning
- Showcase basic video gear and explain their benefits and limitations
- Illustrate different ways to stage talent
- Discuss how to work with inexperienced talent

At its core, production simply means recording sound and images that someone else will want to watch.

ANNABELLE'S TEXT ARRIVED just as I finished writing the chapter on photography. "Help," she wrote. "My boss wants to record a video for Facebook. What should I do?"

"Shoot it with your cell phone," I replied. "But, if she wants it to look good, shoot horizontal video. Put your phone on a tripod so you get a steady picture. Record your audio using a lavalier mic; don't use the camera mic. Put your talent near a big, north-facing window, without getting the window in the shot. Get a big piece of white foam core card to use as a reflector to smooth out the shadows. Finally, have them sit at a conference table and turn their body slightly so they don't face the camera square on."

She took my advice and spent about $150 for a small tripod for her phone and a lavalier mic. The results looked and sounded great.

Video production turns ideas into images. While it doesn't need to be expensive to look good, video production is far more than simply pointing a camera and pressing *Record*.

When I first started shooting interviews, it took a crew of four and a quarter of a million dollars of video gear to record it. Now, any of us can create video with our smartphones. That's part of the reason why I'm writing this book;

Fond Memories

Thinking about Annabelle's production with one person running camera, another holding a white card, and a third person speaking on camera reminds me of one of my favorite productions.

Many years ago, I was the co-producer and director of the *Baltimore Children's Festival*, a live, three-hour production in downtown Baltimore for WJZ-TV. Our host was a virtually unknown local talk show host named Oprah Winfrey.

We had a crew of 85 spread across six buildings throughout downtown Baltimore, a cast of 12 middle-school kids, more than a dozen live performances, and our own animated version of Sponge Bob Square Pants. It took three months to plan. It was one of the most fun projects I ever directed, and that show was on Oprah's demo reel when she was recruited, a few years later, to *A.M. Chicago*.

One of the cameras for the *Baltimore Children's Festival*. Cameras looked a bit different back then.

video production for large audiences is no longer limited to large crews with large budgets.

To create persuasive video today, we need to consider the holy trinity of video production: cameras, lights, and sound. Video production spans all levels of complexity, and, with today's gear, low budgets can still create high quality. That's what this chapter is about.

Planning Your Production

The biggest challenge in any new project is learning what you don't know. But, when it comes to shooting video, you already know enough to start.

At the start, I want to stress: Know your deliverables! Deliverables are the ultimate format for your finished project. These are the master files you need to deliver to your client, boss, or the world. Sometimes, the specs for what you need to provide are pretty vague; other times, they are very specific. Know what you need to deliver; then, as much as possible, shoot that format.

Key technical specs for deliverables include:

- Horizontal or vertical video
- Frame size (measured in pixels)
- Frame rate (this is the hardest spec to convert)
- Codec

It is easy, in production, to set any of these options. However, if you make a mistake in production, it is often difficult later in editing to convert them.

Help! I Need to Shoot a Video

Production is the process of recording sound and images. It fits between pre-production (planning) and post-production (editing). And, while production can be enormously complex and expensive, at its core, production simply means recording sound and images that someone else will want to watch. For Annabelle's boss, a simple, straightforward talk to the camera is fine. Hollywood blockbusters are more complex.

Remember, video is a collaborative act. You don't need to know all the answers. Ask your crew—and talent—for suggestions. Do a test shoot with friends to discover what you don't know; rehearsal is key. And, if this is your first video project and your career is on the line, hire a pro to advise you. There's no harm in saying, "I don't know this—teach me."

To create persuasive video today, we need to consider the holy trinity of video production: cameras, lights, and sound.

Murphy's Law Rules

As you begin shooting, there are five things to keep in mind about media:

- Plan for chaos.
- Review all shots before changing setups.
- Copy all media from the camera card to your storage.
- Have a backup plan—duplicate all media as soon as you shoot it.
- Organize the media so you can find it later.

Murphy's law rules production: "Whatever can go wrong will go wrong and at the worst possible time."

Murphy's law rules production: "Whatever can go wrong will go wrong and at the worst possible time." The crew is late, actors can't remember lines, someone forgot to get parking passes, the lens won't attach to the camera, a light broke and there are no replacement bulbs...whatever can cause the most inconvenience will undoubtably occur.

It is easy to say "plan for it," but that doesn't help my ulcer when an expensive crew is sitting around during a too-short production day waiting for a converter cable to arrive from Best Buy, which happened to me last year.

The next stress point is what to do once your scene is shot. Before you do anything, review your recordings. Is there picture? Is the picture in focus? Is there audio? Is the audio loud enough?

I can't tell you how many great takes were ruined because of a technical glitch. If there's a problem and you discover it on set, it is easy to reshoot. If you don't discover it until editing, well, your options are much more limited. Verify your media before you move on to the next shot.

As soon as possible, depending upon what you are using for a camera, once you verify the shot, copy the media to at least two storage devices. Making copies is always a good idea; Murphy is always lurking right around the corner.

Media management is an essential part of all digital media.

Basic Camera Gear

Next, I want to illustrate different camera gear, their strengths and weaknesses, and suggest which to use for your next project. There are five broad categories of cameras, each with advantages and limitations.

- Mobile
- Video
- DSLR
- Cinema
- Drones

Mobile Devices

As debates rage about what is the "best" camera, this old truism comes to mind: The best camera is the one you have with you. For most of us, that means a phone (**FIGURE 13.1**).

Mobile phones and tablets are ubiquitous with ever-improving cameras, but for all their popularity, mobile devices also have limitations (**TABLE 13.1**).

FIGURE 13.1 An Apple iPhone. *(Image courtesy of Apple, Inc.)*

TABLE 13.1 Mobile Devices as Cameras	
ADVANTAGES	**LIMITATIONS**
Many now record 4K video.	The default lenses are wide angle.
Many can take multi-megapixel still images.	Too often, depth of field is deep, meaning everything is in focus.
Image software continues to improve.	Additional lens selection is limited and awkward to change.
Skin tones are lovely.	
Virtually the entire image is in focus.	Zooming into close-ups tends to create blurry or pixelated shots.
	Camera controls are automatic, with limited variation in settings, though lots of special visual effects.
	Mobile devices have limited audio recording capability.

One of our goals is to control the images we create. However, with a mobile device we trade ease of use for a lack of control. Mobile devices are ubiquitous, but they are not always the best choice. Here are other camera options to consider that are specifically designed for shooting video.

Video Cameras

Video cameras (**FIGURE 13.2**) are the first step up from a mobile device. Designed for single-person operation, requiring almost no additional gear, these cameras are designed for self-contained operation. These cost about the same as DSLR cameras and generally record to MicroSD cards stored in the camera. **TABLE 13.2** lists other advantages as well as some limitations.

Manufacturers of this style of camera include JVC, Canon, Panasonic, Sony, and others.

FIGURE 13.2 This is a video camera, with integrated lens and audio recording. *(Image courtesy of JVC.)*

TABLE 13.2 Video Cameras

ADVANTAGES	LIMITATIONS
Integrated lens smoothly zooms from wide shots to close-ups	Cannot change lenses
	Optimized for video, not still images
Built-in microphone and audio support	Records a highly compressed video format
Ability to record at least two mics at once, with good to excellent audio circuits	Less image control than a DSLR
Support for different video recording formats and image settings	
More image control than a mobile device	

DSLR Is a Noun

DSLR (digital single lens reflex) describes a technology for the original format of this camera. It has morphed into a noun that describes a range of cameras that are designed to shoot high-quality still images that also shoot video. Some prefer to call these cameras interchangeable lens cameras (ILCs). In this book, I'll refer to them as DSLR, because it is a name most of us are familiar with, even though the technology has grown.

FIGURE 13.3 This is a DSLR camera. Lenses are sold separately. *(Image courtesy of Canon, Inc.)*

DSLR Cameras

The next step up from a video camera is a digital single-lens reflex (DSLR) camera which refers to a specific camera technology (**FIGURE 13.3** and the sidebar "DSLR Is a Noun.")

Originally designed for high-quality still images, these cameras have expanded to support both stills and video (**TABLE 13.3**). These come in two flavors: mirror and mirrorless. Mirrorless is the latest technology, reducing the size and weight of a camera, as well as most of the moving parts inside. This improves weight, image stability, and monitoring. DSLR cameras, when used for video, tend to require more time for setup, framing, and focusing because of limited monitoring and the shallow depth of field that most DSLR lenses provide.

TABLE 13.3 DSLR Cameras	
ADVANTAGES	**LIMITATIONS**
The vast array of interchangeable lenses. It is impossible to overstate the benefits of picking and choosing specific lenses for a camera.	The camera body is sold separately from the lens.
	They cost more than mobile devices.
Shoots high-quality video and stills.	They have very poor audio recording, generally requiring recording audio on digital audio recorders.
Greater control over depth of field.	
Lower cost than higher-end cinema cameras.	Many overheat when recording video for more than 20 minutes.
Greater control over the image.	They are not designed for video shoots requiring fast setup, quick shooting, or long recording sessions.
The ability to record to external storage supporting higher-quality images.	
	Monitoring and focusing during video recording are often problematic.
	They require a lot of outboard equipment to work well recording video.

The biggest challenge in any new project is learning what you don't know.

DSLR cameras are also popular with podcasts due to their small size, affordability, and great images. Manufacturers include Canon, Nikon, Sony, Blackmagic Design, and others.

Cinema Cameras

At the high end of the food chain are the cinema cameras, as illustrated in **FIGURE 13.4**. Manufacturers include Arri, Sony, RED, Panasonic, Blackmagic Design, and others.

These highly-configurable cameras require an investment, but they make spectacular pictures (**TABLE 13.4**).

FIGURE 13.4 Cinema cameras are the high end of media production. They provide the highest image quality and are infinitely configurable. *(Image courtesy of Arri, Inc.)*

TABLE 13.4 Cinema Cameras	
ADVANTAGES	**LIMITATIONS**
Highly configurable cameras	These cameras are seriously not cheap
Maximum control over the image	They require an experienced crew to operate
Support for a wide variety of lenses and codecs	They require recording audio to separate digital audio recorders
Used every day in recording high-end feature films, television programs, and commercials	They require a lot of outboard gear to take full advantage of their features

FIGURE 13.5 Drones combine mobility with high image quality but without audio. *(Image courtesy of DJI.)*

Drones

Drones are a special category of camera, as shown in **FIGURE 13.5**. These combine the mobility of flight with a high-quality camera (**TABLE 13.5**). Increasing governmental regulations governing who can fly these and where they can be flown has reduced much of the reckless behavior that appeared when drones were introduced. The dominant drone manufacturer is DJI, while other companies include Yuneed, UVify, and Parrot.

Which camera should you pick? It depends upon your project. For Annabelle, the ease of use of an iPhone met her needs. If you need to move quickly with better audio and high-quality close-ups, a video camera is a better choice. For more control over the image, a DSLR provides more options. And, if you have the budget for high-end gear and crew, the superb images cinema cameras create are unparalleled.

TABLE 13.5 Drone Cameras	
ADVANTAGES	**LIMITATIONS**
Excellent for traveling wide shots in otherwise inaccessible locales.	No ability to record audio.
	Noisy propellers.
High-quality images.	Not suitable for small rooms or close-ups.
	Extensive governmental regulations on operation and flight.
	Limited flight time due to small batteries.
	Often record to a highly-compressed video format.

Camera Support Gear

Before leaving the discussion of hardware, there are four other devices you should consider adding to your inventory: tripods, tripod heads, sliders, and gimbals. All of these help support your camera.

Tripods

Cell phones are easy to hold in your hand, so are many video cameras. The problem is that your hands are not very steady. As frame sizes increase and your videos are displayed on a large screen, excessive camera shake can cause motion sickness in viewers. At the least, too much shaking is *really* distracting. The *last* thing you want is someone throwing up in the middle of watching your video.

If all you need to do is hold the camera steady, any tripod will do. The trick is when you need to move the camera smoothly while recording. Then, the solidity of the tripod makes the difference between a shot you can use and one you can't. **FIGURE 13.6** illustrates one of many tripod designs.

There are *many* tripod manufacturers, including Libec, Manfrotto, and Sachtler. A cool version for light cameras is the Joby Gorilla-Pod, a flexible stand and grip that holds a camera steady in situations where a typical tripod won't fit.

As you decide what to buy, look for a tripod strong enough to hold the weight of your camera. If you plan to carry the tripod a lot, carbon fiber decreases weight but increases cost. If you plan to do lots of pans and tilts, a "spreader," located in the middle of the tripod, keeps the legs from destabilizing during camera moves.

Tripod Heads

A tripod head attaches the camera to the tripod. **FIGURE 13.7** illustrates two different tripod heads. If you are shooting only still images, a simple ball head will do. But if you plan to move the camera while recording, a fluid head becomes essential. This technology allows you to smoothly tilt a camera up and down, or pan from side to side, without jerkiness.

Regardless of which style tripod head you buy, make sure it is strong enough to hold the weight of the camera. Manufacturers include Libec, Manfrotto, Sachtler, and many others.

FIGURE 13.6 A tripod provides a stable platform to avoid distracting camera shake. *(Image courtesy of Manfrotto.)*

FIGURE 13.7 A tripod "head" attaches the camera to the tripod. A ball head (top) is good for stills, while a fluid head (bottom) is best for video. *(Images courtesy of Manfrotto (top) and Sachtler (bottom).)*

Sliders

Since the reason we shoot video is to attract a viewer with movement, a slider can make camera movement easier. Many times, it is easier to stage and light stationary talent while moving the camera gently. A slider makes this easy (**FIGURE 13.8**). This is a bar perched on a tripod that holds the camera steady while allowing it to smoothly slide from one end of the bar to the other. It always surprises me how just a bit of movement makes a big difference in how we view a scene. Putting something like a plant or candle in the foreground, then slowly sliding the camera past to reveal two people talking in the background is an effective technique to draw attention to the conversation.

FIGURE 13.8 A slider, mounted to the top of this tripod, smoothly slides the camera to create a more visually interesting shot. *(Image courtesy of Libec.)*

Gimbals

Just as drones provide mobility for a camera in the air, gimbals provide mobility for a camera on the ground. As illustrated in **FIGURE 13.9**, these stabilize a handheld shot to minimize side-to-side movement (though they don't help with up/down movement, which means that a gimbal operator needs to learn a new way to walk smoothly).

While not as stable as a tripod, gimbals allow cameras to follow talent in a way that no other technology supports. Manufacturers include GoPro, Glidecam, DJI, and others.

The goal with all of this gear is to provide solid support for your camera so that your audience can concentrate on your story, instead of wondering whether they need to make a quick trip to the sink.

FIGURE 13.9 A gimbal, like this Ronin-S, stabilizes a handheld camera as the operator moves it through space. *(Image courtesy of DJI.)*

Renting: An Alternative to Buying

If you plan to create a lot of videos, your best option is to purchase the gear you need. That way, it's always available, and, more importantly, you know how to use it.

But, if you do only the occasional project or you are trying to decide what gear to buy, an excellent option is to rent the cameras you need. This provides an ideal opportunity to test the gear under normal production conditions to see if it does what you need it to do.

The biggest benefit, to me, is that you can rent gear that you may not be able to afford to buy. As well, this also allows you to work with the latest production technology.

As I mentioned in our discussion of audio, this is exactly what I did to find the right microphones for my studio. I rented ten different mics, brought in the talent I was planning to use frequently, and asked my co-workers to join me in evaluating these mics. What surprised me was how different the mics sounded and how consistent my listening team was in their opinions. I'm still using those same mics many years later.

In the shots that follow, though I don't often mention it, look at how each shot is framed to follow the Rule of Thirds.

The disadvantages to renting are that you are always learning new gear, you don't own it when the project is done, and it may not be available when you need it.

Talent Staging

Once we have the gear we need and the people to run it, it's time to work with the talent. Let's take a look at a variety of ways to stage talent. Even though cameras and/or talent are moving, the Six Priorities and the Rule of Thirds still apply to guide where the viewer should look throughout your video. For example, carefully look at how your talent sits in relation to the camera. While this scene is set in a bar, it holds equally true for interviews.

Posing

Compare the difference posing makes between the two shots in **FIGURE 13.10**. Turning the talent's body so they are slightly profile to the camera makes for a more visually interesting shot. This still allows them to look directly at the camera, if you want, but the turn looks better.

FIGURE 13.10 Compare the difference between the talent squarely facing the camera (top) versus a slight turn (bottom). The turn looks better.

I use this technique all the time. If the talent is sitting, I ask them to point their knees about 30° away from the camera lens even if this means physically turning the chair so they are sitting at the correct angle. If they are standing, I'll ask them to slightly rotate their body.

All too often, both close-ups and wide shots are taken from the same position. This is a mistake.

Finding the Best Close-Ups

Not everything we shoot will be an interview, which is also called a "talking head." (Why? Because it's a close-up of a head, talking.) In Chapter 7, "Persuasive Photos," we looked at the 180° Rule, as shown in **FIGURE 13.11**, an imaginary line connecting the noses of the two principle speakers in a conversation. This rule was actually invented decades ago for film and video because most inexperienced camera operators put the camera in the wrong spot.

As we learned earlier, when shooting a conversation, all cameras need to stay on the same side of the 180° line. What is often overlooked, however, is that the best close-ups come from getting close to the line, while the best wide shots are from the center. All too often, both close-ups and wide shots are taken from the same position. This is a mistake.

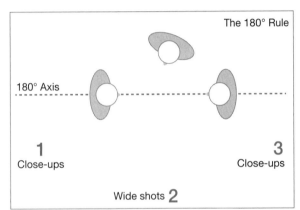

FIGURE 13.11 The 180° Rule defines a line connecting the two principle actors. Different camera positions are used for specific shots.

In **FIGURE 13.12**, Kim, on the left, is chatting with Allison. The camera is centered between them. Kim is looking camera right, at Allison. Allison is looking camera left. Note how, even though they are talking to each other, both actors are turned slightly out, opening their profile to the camera. (This matches the Camera 2 position in the floor plan shown in Figure 13.11.)

FIGURE 13.12 Kim and Allison are having a conversation.

FIGURE 13.13 The camera moves over Kim's shoulder for a closer shot of Allison, who is still looking camera left.

FIGURE 13.14 We move behind Allison, again near the line, to get a good close-up of Kim.

FIGURE 13.13 shows what happens if we move the camera closer to the line on the left. From this position we can get an over-the-shoulder shot of Kim to Allison, providing a nice close-up of Allison's face. Just as in the wide shot, Allison is looking camera left. (This matches the Camera 1 position in the floor plan shown in Figure 13.11.)

See how the lighting reinforces screen direction, where one side of both their faces is brighter than the other?

Switching sides (**FIGURE 13.14**), we move the camera over Allison's shoulder to get a good close-up of Kim, who is looking camera right. This angle also matches the screen positions of the opening wide shot. By staying close to, yet on the same side of the 180° line, we get interesting, tight close-ups without confusing the screen direction. (This matches the Camera 3 position in the floor plan shown in Figure 13.11.)

But as soon as we cross the line, as shown in **FIGURE 13.15**, we confuse the audience because now Kim is looking in the wrong direction, implying she and Kim are no longer talking to each other. Kim and Allison can't both be looking screen left and still have a pleasant conversation.

There are times where you *want* to confuse the audience. Action sequences, chases, and arguments are examples of this. But, most of the time, you want to honor screen direction from wide shots to close-ups.

Let's shoot another dialogue, this time an argument, and see how we can increase the emotional level by changing the camera position, framing, and focus.

FIGURE 13.15 Crossing the 180° line means Kim is looking in the wrong direction and the lighting doesn't match.

FIGURE 13.16 The wide shot sets the scene: where the actors are and where they are looking.

FIGURE 13.17 Move closer to emphasize faces, while losing focus on the background.

FIGURE 13.18 Move in really tight, focus on Allison, and have her look out of frame. This puts her emotions at the center of this scene.

In this series of three shots, we are not zooming in. We are physically moving the camera. As we do, the background changes and defocuses. Remember, the eye goes to that which is in focus. Sometimes you should zoom. Other times, you should move the camera. There's a difference.

In **FIGURE 13.16** Allison and Kim are having an argument. How do we know? They are not looking at each other. The camera is wide, showing the environment and the position of both women. Let's boost the emotional level and tell the audience who is at the center of this scene by moving closer and adjusting the depth of field.

As we move closer in **FIGURE 13.17,** the background fades into a soft blur, which is why I like to use depth of field; it directs the eye. The tighter shot also provides a much better view of the emotions in the actor's faces.

As we move in *really* close, as shown in **FIGURE 13.18,** Allison is the only one in focus, as well as the largest object in the frame. Our eye *has* to go to her first. We also emphasize her rejection by having her look into the short side of the frame. This creates a constricted, trapped, and alienated feeling, reinforcing the emotional power of the scene.

FIGURE 13.19 Wide, dark, and lonely—emphasizing her isolation.

Not all compelling images need to show smiling people. Our goal is to attract the eye. Emotions are key to that attraction.

Here's another emotionally charged option: loneliness. Look at how the distance, lighting, and emptiness in the frame shown in **FIGURE 13.19** emphasize Allison's loneliness and isolation. The shot is wide, her face is dark, and the room is empty. Even so, this is still framed according to the Rule of Thirds.

Staging Entrances with Impact

Another staging technique that gives us an opportunity to tell the audience where to pay attention is in entrances and exits. In real life, people enter or exit a room from anywhere there's a door. In theater, most entrances are horizontal, meaning the actor walks parallel to the front of the stage. In video, though, entrances and exits move to and from the camera.

In **FIGURE 13.20**, both Allison and Kim are the same size. Kim is brighter, but Allison is moving. There's no clear telegraphing of where the eye is supposed to look first. In fact, because Allison is so dark, we tend to see Kim, on the right, as the focus of this scene. That may be our intent, but, often, it's because we didn't think about the entrance clearly enough.

FIGURE 13.20 Allison enters the frame on the left and walks to the bar.

Movement that starts or ends close to the camera is much more compelling, as you can see in **FIGURE 13.21**. The actor changes size as they move, which, in addition to their movement, is more compelling than simply moving sideways through space. In this example, Allison enters next to the camera on the camera right side. She is larger and brighter than anything else in the frame. She then changes size as she walks closer to the bar to begin a conversation with Kim, which draws the viewer's eye into the scene.

FIGURE 13.21 A costume change and repositioning Allison to enter next to the camera makes this shot much more effective.

Props: Something for Talent to Work With

Notice that in both Figure 13.20 and Figure 13.21, Kim is holding a prop, a cloth, as she wipes the bar. Props are a handy thing to give to talent. It gives them something to do with their hands, and it gives them something to focus on.

Here's another example. In **FIGURE 13.22**, notice the difference a simple prop makes. Yes, an actor can make do with nothing. But they can do a whole lot more when they have something to work with. Props provide a focus for a scene and, often, an extra layer of meaning.

Props provide a focus for a scene and, often, an extra layer of meaning.

All these staging ideas are simple, don't cost any money, and can be used in a wide variety of situations, from interviews to commercials to drama. Remember, the entire environment of the shot—camera, lights, sound, talent, set, props—combines to reinforce your message to the audience.

Working with Inexperienced Talent

We've talked a lot throughout this book on how to work with talent. If you are shooting your first video, keep things simple: Have the people on camera hold still. That makes them easier to light. Lighting people who are walking around is much more difficult.

The key to working with inexperienced talent is reassurance.

Inexperienced talent ranges from nervous to scared. They are doing something out of the ordinary, and there's a substantial amount of fear associated with it. They don't want to make a fool of themselves. Speaking to a camera is nerve-wracking; I've been on camera a lot, and it still makes me nervous.

The key to working with inexperienced talent is reassurance. There's no crime in being nervous. The trick is to give them something to focus on that isn't the camera. That's generally either me or a fellow actor.

Our goal is to make talent look good on camera. Most of the time, that rule also means they are not particularly comfortable. Soft, cushy chairs may feel great, but anyone sitting in them looks folded and sunken. It is much better to put your talent on a hard chair or stool. This forces better posture and, by being uncomfortable, gives them something to think about other than how nervous they are to be on camera.

In interviews I tell guests to look at me and ignore the crew and gear. They can't, of course, but giving them someone to look at helps to calm them. I also show them around the set when they first arrive so they aren't left wondering what's going on. However, after the tour, I get them settled and turned so that they see as little of the gear as possible.

(That's assuming it's a production where I'll be near during their performance. If I'm in a control room or off set, I make sure the stage manager has instructions to keep them relaxed.)

I also constantly praise their work, even if they don't deserve it. They are doing the best they can, and if it doesn't meet my standards, it isn't their fault—I was the one who decided to put them on camera in the first place. It's similar to sports. The athlete wants to know: "How am I doing, coach?" My job is to coach them into the performance/interview/presentation I need.

One other thing. Assign someone, ideally hair and makeup, to take one last look at them before cameras roll and reassure them that they look fine.

> ### Nothing Beats Rehearsal
>
> Nothing beats rehearsal. Before any talent walks onto the set, connect and test your gear. There is no worse feeling than recording a great scene, or conducting a great interview, only to discover that a tiny piece of gear wasn't working and everything was lost.
>
> Yes, I've been there. It is not pleasant.
>
> Production is pressure-filled and time-limited. Test your gear. Rehearse your talent. Practice your moves. Do everything you can to make sure that, this time, Murphy stays far away.

Production is pressure-filled and time-limited. Test your gear. Rehearse your talent. Practice your moves. Keep Murphy at bay.

KEY POINTS

This chapter built on elements we covered in earlier chapters, because production is part of the process extending from the initial planning (pre-production) to editing (post-production). Here are the key points from this chapter:

- Every video tells a story, not just what the talent say but how the scene looks.

- Everything we learned about persuasive still images applies to video with the added benefit that, in video, everything moves.

- Design your production based upon your deliverables.

- The three major technical groups in production are cameras, lighting, and audio (sound); though, clearly, everyone from production design to makeup plays an important role.

- Five major camera groups are smartphones, video cameras, DSLRs, cinema, and drones. Each group is designed for different types of production.

- If this is your first major project, consider renting your gear and hiring pros to make it look great.
- When working with talent, staging, blocking, and props help them do their best on camera.
- Practice with your gear, rehearse your talent, and allow for problems.
- The key to succeeding with inexperienced talent is reassurance.

PRACTICE PERSUASION

Shoot a conversation between two friends or family members. Follow the 180° Rule; then break the rule and see the difference between staying on one side of the line versus crossing it.

Which position allows the audience to figure out what's going on and which confuses it? When would you want to confuse an audience?

PERSUASION P-O-V

A LITTLE KNOWLEDGE IS A DANGEROUS THING

Larry Jordan

I left graduate school at the University of Wisconsin at Madison, where I studied Radio/TV and Film, to take a job at the only two-camera television station in the state of Montana: KGVO-TV. (In those days, the other TV stations in Montana had only one camera. This was a long time ago, and studio cameras were not cheap.)

Based in Missoula, I was hired to direct the evening news. But, truthfully, I was also a street reporter, production staffer, and the only person who regularly shot news film with audio; the rest of the staff just shot silent film, because it was easier to work with.

Fresh out of grad school, shooting and directing at a broadcast TV station— I was in hog heaven!

One day, the local drug store asked us to create a television commercial for their store featuring Chanel No.5 perfume. I volunteered to help our production manager light the set because I studied lighting in college and "knew all about it."

In those days, an RCA TK-43 studio television camera stood over 5 feet tall, was 6 feet long, weighed close to 1,000 pounds, and was mounted on a three-wheel rolling studio pedestal. (I once terrified the news anchor by rolling the camera in for a close-up, only to lose control and have it smash into the set. Those cameras had a lot of mass!)

Another attribute of those early TV cameras was that they took a *lot* of light. The studio lights we used were measured in thousands of watts!

To create this ad, the drug store gave us a large glass Chanel No.5 bottle filled with a colored alcohol mixture as a prop. Fluids look best if they are lit from underneath, as well as using the three-point lighting I talked about in Chapter 7. So, I put the bottle on a sheet of plate glass, with a studio light under it. I then added key, fill, and backlights hung from the lighting grid in the ceiling. Each light was 1,000 to 2,000 watts, but they were at varying distances and heights from the bottle, which allowed me to control the intensity and shadows.

Image Credit:
Chanel website

I left the studio and headed to Master Control to load a tape on the video tape machine to record the shot. As I looked at the monitor, the perfume looked spectacular! The glass sparkled with highlights, the perfume radiated a golden glow, the scene had a lovely high-quality look.

I hung a tape on the VTR, and as I pressed Record, I looked back at the monitor—and saw nothing. The screen was black.

Concerned, I went back into the studio to discover Dave McLean, our production manager, beating at a 12-foot column of flame with a broomstick! I dashed over to the fire extinguisher, and we quickly put out the fire.

In my enthusiasm for lighting the perfume, I put the plate glass that held the perfume directly on the 1,500 watt scoop light. The heat from the light cracked the glass, the glass bottle fell and broke, spraying the fake perfume solution directly on the shattered 3,200° tungsten bulb, causing the solution to explode.

The worst part, after everyone got done yelling, was that the smell in the studio was so bad we had to do the news that night from the parking lot. And it was January.

Clearly, a little less enthusiasm and a bit more planning would have yielded better results. (The image did look great, though. Before the explosion, that is....)

If chess has any relationship to film-making, it would be in the way it helps you develop patience and discipline in choosing between alternatives at a time when an impulsive decision seems very attractive.[1]

STANLEY KUBRICK
FILMMAKER

CHAPTER 14

VIDEO POST-PRODUCTION

CHAPTER GOALS

The goals for this chapter are to showcase an effective editing workflow and editing techniques that allow you to take the images captured during production and turn them into a program that someone else will want to watch.

The goals of this chapter are to:

- Share a workflow for editing that applies to all editing systems and visual storytelling
- Explore each step of the workflow as a way to learn basic editing
- Discover how to use Apple Final Cut Pro X for storytelling
- Learn what to consider when editing any project

The purpose of editing is to create a clear, compelling, and engaging message from the elements shot during production.

THE EVIDENCE WAS ALL THERE: super-fast, unjustified switching between shots coupled with overwhelming and distracting visual effects. The verdict: an insecure editor. Fast cutting in an action sequence makes sense; fast cutting in an interview does not. The old adage applies: "Just because you *can* does not mean you *should*."

The purpose of editing is to create a clear, compelling, and engaging message from the elements shot during production. To accomplish this, an editor has two options: one, *showcase* the talent on camera or, two, *hide* the fact that the person on camera has no talent. Either way we still need to create a clear, compelling, and engaging message.

What Is "Good" Editing?

As you might guess, editing is assembling a series of video clips together to tell a story. But what is "good editing"?

As I was researching this chapter, I did a Google search for "bad video editing." I found lots of different lists and discovered that errors consistently fell into three groups (**TABLE 14.1**).

TABLE 14.1 Indicators of Poor Editing		
AUDIO	**NOT SEEING WHAT'S ON THE SCREEN**	**CONTENT**
Poor audio mix.	Flash frames caused by edits where extra frames of old clips are left at an edit point.	Pacing is too fast.
Audio is out of sync with video.		Pacing is too slow.
	Jump cuts. Unjustified, instantaneous shifts in position.	Holding shots too long or too short.
The emotions in the music don't match the tone of the video.	Bad transitions. Images changing in the middle of a dissolve.	Keeping shots that are lovely but don't belong.
	Inconsistent graphics/fonts/styling.	
	Poor camera framing.	
	Poor color.	

It was interesting to me that poor audio made the "top three" problems on every list I reviewed. Audio is really that important to video.

It's a depressing list. However, these problems can all be fixed by following a simple precept: "Slow down and pay attention." Your ears will tell you if the audio mix doesn't sound right. Concentrate during playback and actually see what's on the screen, as opposed to what you *expect* to see on the screen. Finally, ask yourself: Does my story make sense? Am I telling my story too quickly? Am I distracting the audience with images or sound that don't relate to my story?

Slow down and pay attention—you are spending days, even weeks, assembling a project. Your audience gets to see it only once. Do everything you can to share your story clearly with an audience that may not be paying full attention.

"Slow down and pay attention" fixes most problems.

There are lots of excellent resources on the web that can help you improve. In addition to my website (LarryJordan.com), there are blogs at PremiumBeat.com, MotionArray.com, Videomaker.com, and Pond5.com, as well as training at NoFilmSchool.com, Moviola.com, and many others.

See What's There

Over the years, my students have proven that when they are editing video, they see what they expect to see, not what's actually on the screen. I don't know an easy way to correct this except to point out—as I did in Figure 3.4 in Chapter 3, "Persuasive Writing"—the Google goggles. Not seeing what's right in front of you is an easy trap to fall into.

Find the Rhythm

A key editing concept is "rhythm." Speech is rhythmic. Music is rhythmic. Movement is rhythmic. We respond to rhythm, and we need to reflect that rhythm in our editing, not just in the dialogue, but in the timing, pacing, and sequencing of our shots.

When Should You Cut?

Walter Murch, Oscar-winning editor of *Apocalypse Now*, *The English Patient*, and many other major feature films, spent 100 pages in his outstanding book "In the Blink of an Eye" discussing where to cut between two shots. The short answer is: Edit when your talent blinks.

As we edit, we use pictures to show what is happening, dialogue to tell the story, sound effects to make the pictures believable, and music to drive the emotion. Nothing appears in a video by accident. You may not have planned for that prop to break, for that ad-libbed line, or for an interesting juxtaposition of shots that caught your attention during the edit—but, once you put it on the Timeline, it stays there because you *want* it there.

I've written eight books, so far, about editing. There's no way in a chapter I can teach you everything you need to learn. Instead, I want to share with you a way to think about editing that will make learning and using any editing software faster and easier.

I frequently refer to Apple Final Cut Pro X as "Final Cut" and "FCP X." Also, just so you know, it's pronounced "Final Cut Pro Ten."

To do this, I'll use examples from Apple Final Cut Pro X. Why? Because years of teaching Final Cut and Adobe Premiere Pro and working with other editing software have shown me that Final Cut Pro X is faster and easier for noneditors to learn and use successfully. Apple, Adobe, Avid, Blackmagic Design, and others make professional-grade software that can edit everything from home movies to major motion pictures. They all do professional-level work.

Will other editing software work for your projects? Most likely, yes. Will it deliver excellent results? Again, probably yes. Pick the tool that works best for you. For this book, that tool is Apple Final Cut Pro X.

I often use "cutting" as a synonym for editing, because editing began first with film that was cut with a razor, then glued back together.

Another reason for learning to cut using professional tools is that when the chips are down and deadlines are looming, you need software that has the depth of features and stability to get your work done. If all you need to edit is a home movie, any editing software will do. But, when money and reputation are at stake, you need software that, when you really lean on it, it supports you.

With very few exceptions, video editing is the most technologically challenging task you can do on a computer. It requires a high-power system, especially as codecs get more and more efficient at compressing media. Editing depends on the tightly integrated efforts of the CPU, GPU, RAM, storage, operating system, software...every component in your computer is challenged during video editing. Editing may appear easy, but, under the hood of your computer, there is a *lot* going on.

As the editor, you need to balance creativity with technology.

Editing is also a highly creative task because there is an unlimited number of ways to tell the same story. Even in adjusting where clips touch and how we change from one shot to the other, hours can be spent deciding the best place to make the change.

Beyond the creativity and the technology, editing is also a highly logical task. Organizing the story, tracking clips, managing media, meeting deadlines...all of these different challenges are consolidated into one activity: editing.

As the editor, you need to balance creativity with technology. You need to look at the story you are telling from the audience's point of view: create a rhythm and flow to your story. Make sure there aren't any jarring edits, jump cuts, or visual distractions that take viewers out of your story.

This chapter looks at the process of editing. Along the way, I'll provide some tips on the creative side, but, like most creative activity, the best way to learn something is to do it, watch it, critique it, then try again.

An Editing Workflow

There is never enough time—not in planning, not in production, and definitely not in editing. No creative task is ever completed; instead, they are abandoned as time runs out. This means we need to be as efficient with our time as possible and that means we need a workflow.

Over the years, I've developed a 12-step editorial workflow that answers the question, "What should I be doing *right now*?" When I teach high-school students, they first want to learn to create effects. When I teach college students, they first want to learn to edit.

Where we *really* need to start, though, is planning. That's where this editorial workflow fits; it keeps all our energy and enthusiasm focused. This chapter is organized around a 12-step workflow presented in **TABLE 14.2**.

TABLE 14.2 The 12-step Editorial Workflow

CREATE THE STORY	POLISH THE STORY
1. Plan the project.	7. Add transitions.
2. Gather the media.	8. Add text and effects.
3. Organize the media.	9. Create the final audio mix.
4. Build the story.	10. Finalize the look and colors.
5. Organize the story in the Timeline.	11. Output the project.
6. Trim the story.	12. Archive the project.

Chapter 10, "Video Pre-production," defined media management as organizing clips by camera card and source. Once we move into editing, organization shifts to subject and scene. For most edits, how we organize clips in our video editing software does not match how clips are stored on our hard disk.

Define Terms

Before we jump into the process of editing, there are five Final Cut terms I need to define:

- **Library.** This is the master container Final Cut uses for your edit. It holds the information necessary to find your media and how it is edited. While it is an icon you see on the desktop, a library is a "super-folder" that acts like a single file but actually holds lots of different stuff.

- **Event.** This is a fancy name for a folder inside Final Cut. Events hold the projects, video, audio, and stills that you need for your edit. There's no limit to how many files you can put in one event. Events are stored inside the library, but you can't store one event inside another.

- **Project.** This holds the instructions about how you want your media edited. Projects are opened in the Timeline. The Timeline is the tool that holds the clips we edit into a project. Timelines can be up to 12 hours long and hold thousands of clips. Projects, like events, are stored inside the library.

I will frequently use the term "hard disk." This refers to all the different storage options connected to your system: internal SSD, internal hard disk, external hard disk, external SSD, external RAID, server volumes...the list seems endless. "Hard disk" is how I refer to all of it.

The script creates the idea, production records the story, and the story is told through editing.

- ■ **Media.** These are the clips that we edit. Clips can contain still images, video, audio, timecode, captions...lots of different elements can be stored in a clip. Media files are huge. Media can be stored inside or outside of the library.
- ■ **ProRes 422.** This is the default codec for editing in Final Cut Pro X. ProRes is an excellent codec family; it was invented by Apple and is used by lots of different companies.

OK, let's get started.

Intermediate Codecs for Editing

We frequently convert media from camera native to an intermediate (or middle) codec for editing. This is done for faster performance and exports, as well as improved image quality for color grading and effects. Final Cut makes this easy using an import option called Optimize Media, which I'll cover in a few pages. Then, during export, we convert our project to a third codec for final distribution.

Step 1: Plan the Project

Unlike many other applications, Final Cut Pro X automatically saves your files the instant you make a change. This means you never need to worry about saving a project as you work.

We covered this extensively in Chapter 10. Here, I simply remind you to understand your message and your audience, to make sure your gear is working and that you know how to use it, and to make sure you know the content, duration, and format to deliver when you are done.

Step 2: Gather the Media

This is part of the production phase, covered in Chapter 13, "Video Production," where we shoot new material and gather existing material from other sources, such as prior projects or stock footage houses, that we need to tell our story.

NONLINEAR EDITOR (NLE) Computer software that can access and edit media clips in any order (nonlinear), unlike earlier film editing systems, which were not computer-based. Another way to think of an NLE is as video editing software, but NLE is faster to write.

This is also the step where the media management we covered in Chapter 10 comes into play. We need to safely transfer the media we shot during production to our storage, make backups, then label the folders so that we know what we shot and where it's stored.

Once our media is safely stored, we need to find specific shots. There are a lot of different systems we can use, from simple paper lists to spreadsheets to media asset management software such as Kyno, Axle.ai, and KeyFlow Pro. It may seem that you could keep all your shots in your head, but, ultimately, you'll discover it will be easier to off-load that media tracking to software. For simple projects, spreadsheets are great; we all know how to use them, and they work.

Step 3: Organize the Media

This step starts the editing process, because clip organization takes place in Final Cut, rather than the Finder.

Continuing the trend of impossibly dark interfaces for media software, **FIGURE 14.1** shows the interface for Apple Final Cut Pro X. While it can support two monitors, it is designed to fit into one. There are four principal sections to the interface:

1. **Libraries sidebar and browser.** This is where we view and organize clips.

2. **Viewer.** This is where we watch clips and projects.

3. **Timeline.** This is where clips are edited into finished projects.

4. **Inspector.** This is where we make changes to clips and effects.

We'll work exclusively with the Browser and Viewer to get clips reviewed and organized. Then, we include the Timeline for editing. Finally, we use the Inspector for effects.

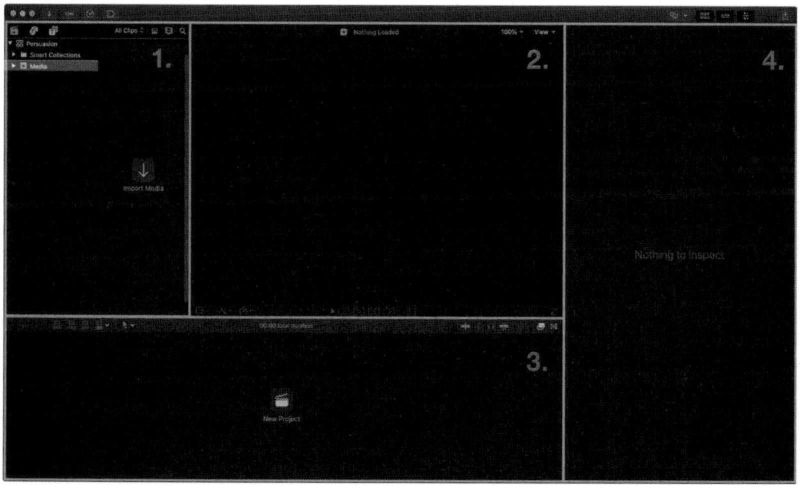

For this chapter, I'm using air show footage courtesy of Hallmark Broadcast Ltd. (HallmarkMediaGroup.com). I'm grateful to them for sharing their clips.

FIGURE 14.1 The Final Cut Pro X interface. 1. Libraries sidebar and browser, 2. Viewer, 3. Timeline, and 4. Inspector.

Import Media

The first step in organizing your clips is to import media. Unlike Photoshop where all the elements for an image are stored in one file, when we import media into any video editing software, we are importing a "link," a pointer, that shows where the media clip is stored in your computer system. Media files are always separate files from the instructions on how they are edited, which Final Cut calls a "project."

Final Cut supports media stored on internal drives, locally attached storage, and servers. However, in most cases, libraries can't be stored on a server. For the rest of this book, we'll just assume that all files are stored locally on external devices. ("External" means hardware that is located outside your computer itself.)

Start Final Cut. If this is your first time running it, it will ask if you want to create a new library or open an existing one. Create a new library, give it a name (I used "Persuasion"), and click Save. By default, it will store these libraries in the Movies folder of your Home directory. This is fine for now. Larger libraries, and media, are best stored on external storage because external devices will, most likely, have far more capacity than your internal drive.

To import media, choose File > Import > Media (Shortcut: Cmd+I). The Media Import window (**FIGURE 14.2**) imports clips and folders that contain clips. (You can also drag clips directly into Final Cut from the desktop, but the Media Import window gives you more control over the process.)

FIGURE 14.2 The Media Import window. Storage devices are listed on the left, media in the middle, and import settings on the right. (Pilot and plane images courtesy of Thalo, LLC / Hallmark Broadcast Ltd. (http://hallmarkmediagroup.com))

Just as an aside, look at the close-up of the pilot in Figure 14.2. It was shot below eye level to give him a heroic look. It's framed according to the Rule of Thirds; he is in focus and is the largest element in the frame. The filmmakers did everything except paint a big red arrow on his forehead saying, "Look here first!"

(Not to mention, the pilot is wearing a flight jacket to boost his credibility and standing in front of a strategically placed jet in the background. This shot was *very* carefully composed.)

On the left in Figure 14.2 are all the storage devices attached to your computer; this would include an iPhone if it were attached to your computer via a cable. In my case, I store all the media I use for training in a folder called "Training Media." (It's similar to the master Source Media folder I discussed in Chapter 10.) Click a hard drive or folder, on the left, to see the contents, which are displayed at the bottom center of this window. Double-click a folder to open it, which is what I did here. I opened the Air Show Footage folder, which has ten clips inside it.

Click any clip in the lower center of the window to select it. Instantly, the entire clip appears in filmstrip view, along with a large image from the start of the clip (**FIGURE 14.3**).

FIGURE 14.3 Selecting a clip displays the entire clip in filmstrip view, plus a larger shot of the frame containing the playhead.

Hover the cursor across the filmstrip to "skim" it. Press the spacebar to play the clip; press the spacebar again to stop. The line you see skittering across the filmstrip is called the "playhead."

This is a fast way to view any clip—select it, skim it, select another.

There are several ways to choose which clips to import:

- Select a single clip.
- Select the first clip in a range, press the Shift key, and select the last clip. All the clips between them are selected.
- Select the first clip then continue selecting any combination of clips while pressing the Cmd key. (This is called "discontinuous selection," a word that I try to work into my conversations as often as possible.)

PLAYHEAD A vertical line that represents the frame currently displayed in the Viewer. The playhead moves in sync with a clip, or project, during playback.

FIGURE 14.4 The import settings are remembered from one import to the next. These are also my typical import settings. The top menu shows the event that media will import into.

TRANSCODE To convert from one media format to another.

■ Click the left-pointing arrow next to the name of the folder (just below the thumbnail on the left). This takes you back up a level. You can then select the entire folder to bring in all the files inside it. There are advantages to bringing in entire folders, which I'll cover when we talk about keywords.

The import settings (**FIGURE 14.4**) on the right of the Media Import window look pretty darn intimidating. But, there are only three we really care about. Once you have them set up, FCP will remember them from one import session to the next. Figure 14.4 also illustrates my standard import settings.

The first choice is selecting the correct event. All media must be stored in an event; remember, an event is just a fancy name for a folder in Final Cut. The event you selected before you opened the Import window is displayed at the top of this panel. You can change it by selecting a different event from this menu. When you first create it, the name for an event will be a date. Don't worry, I'll show you how to change the name shortly.

The second choice is Files:

■ **Copy to library.** This makes a copy of all your media and stores it in the library. This is my recommendation for new editors or editors who need to move the same library from one computer to another. However, storing media in the library makes the library file very big! And it duplicates all your media, which requires still more storage.

■ **Leave files in place.** This leaves files in place on your hard drive, thus saving space. But, it means that if you move your library to another computer, you need to remember to move all your media as well. This is the best choice when sharing media between projects, when sharing media between editors, or when you want to avoid duplicating media.

The third choice is Transcoding:

■ **None.** When nothing is checked, you edit the media shot by your camera, called "camera-native" media. This is the best choice when storage is limited, when you don't plan to take a long time editing the project, and when you don't plan to use many effects or do color correction.

■ **Create optimized media.** This is my recommended choice for most editing projects. This copies and converts your camera-native media to

ProRes 422, which allows for faster rendering and exports, along with higher-quality color and effects. But this option requires more storage space for the transcoded files.

- **Create proxy media.** This is my recommended choice when editing 4K frame sizes or larger or when doing multicamera editing, which Final Cut also supports. I won't cover proxy editing in this book.

For now, with the exception of recommending you use "Copy to library," Figure 14.4 shows my media import recommendations. Use them to get started and modify them as you learn more.

Why We Render Clips

"Render" sounds very technical, but all it means is to "calculate" a new clip from an existing clip. Rendering does not change the source clip. Typical examples include applying effects or color correction to a timeline clip. Final Cut Pro renders in the background to save you time.

Once you have your import settings the way you want, select the clips you want to import then click Import Selected in the lower-right corner of the window. (For this exercise, I moved up one level and imported the entire folder. It's faster than selecting individual clips.)

Configure the Browser

Almost immediately, thumbnails of your clips show up in the browser, as shown in **FIGURE 14.5**.

FIGURE 14.5 After import, thumbnails of each clip appear in the browser (left). The currently selected clip is also displayed in the Viewer (right).

Click the small clip icon, indicated by the top red arrow in **FIGURE 14.6**. The top slider adjusts the thumbnail size. The second slider (middle red arrow) determines how often it displays a thumbnail in the clip. The All setting means one thumbnail per clip. 30s means a new thumbnail every 30 seconds. You can zoom in to a thumbnail every half-second! When waveforms are checked (bottom red arrow), you can see both image and audio on the screen at the same time.

You can change these settings at any time.

FIGURE 14.6 Click the small clip icon at the top (top red arrow) to modify how clips are displayed in the browser.

The list on the far left is the Libraries sidebar. It displays all the open libraries. In **FIGURE 14.7**, I have the Persuasion library open, which has three events: Smart Collections, Media, and Projects. (You can also have multiple libraries open at once.) The Libraries sidebar displays libraries, events, and other elements such as keywords, but not clips. Clips are displayed in the browser.

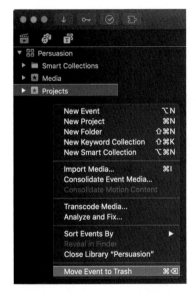

Changing Names

While it is *not* a good idea to change file names in the Finder, it is an excellent idea to change them in the browser, especially if it helps you better understand the contents of a clip. Click the filename to select and change it.

Changing a media file name inside Final Cut does not change the filename of the media file stored on your hard disk.

FIGURE 14.7 Right-click the blank area of the Libraries sidebar for other options.

You can change the names of libraries, events, projects, and media at any time. Click once on the name you want to change to access it for editing. Changing library, event, or project names will change the name of these files on your hard disk. Media names will not change.

More options are available when you right-click something in the Libraries sidebar, as shown in Figure 14.7. Here are some examples:

- To rename a library or event, click its name once to highlight the text then enter a new name. You can use any name you want, but avoid special characters.

- To create a new event, choose File > New > Event (Shortcut: Option+N).

- To delete an event, right-click its name and choose Move Event to Trash.

Favorites and Keywords

Now that our clips are imported, it's time to get them organized. For simple projects, I create two events: Media and Projects. For complex projects, I create more, but I don't usually create a lot. While you can organize media by storing it in different events—and I do—there are two options that make working with media a lot easier: Favorites and Keywords.

A Favorite is simply a clip that you like and that you expect to use in your edit. Click the clip to select it—a gold box appears around it—then press F. The green bar at the top indicates it is a Favorite (**FIGURE 14.8**). To see only favorite clips, go to the All Clips menu (red arrow) and choose Favorites. Favorites are a fast way to flag just the clips you like.

Other options include:

- To hide an unwanted clip in the browser but not remove it from the library, select the clip, and press Delete. The default setting of Hide Rejected in the Clip Filter menu means rejected clips will not be displayed.

- To reveal rejected clips, choose All Clips from the Clip Filter menu.

- To permanently remove a clip from the browser, press Cmd+Delete.

- Type U to remove a Favorite or Rejected flag from a selected clip.

Another way to organize clips, and one I use for all my major projects, is keywords. Keywords, as shown in **FIGURE 14.9**, can be applied automatically, which I did when I imported the Air Show folder. The folder name is applied as a keyword to all the imported clips inside it. (This is an excellent reason for carefully thinking about folder names before importing clips.)

Keywords can also be applied manually, using the floating Keyword panel. To open it, click the key icon at the top, indicated by the top red arrow in Figure 14.9. This allows tagging each clip with relevant keywords in the Keyword panel, shown by the lower red arrow. Use commas to separate each term.

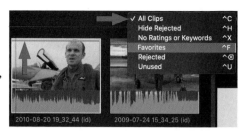

FIGURE 14.8 The Clip Filter menu—which defaults to Hide Rejected—controls which clips are displayed in the browser. Undisplayed clips are not removed, just hidden.

What Cmd+Delete Actually Removes

When you select a clip in the browser and press Cmd+Delete, one of two things will happen:

- If you imported the clip using Copy to Library, the clip in the library will be moved to the Trash in the Finder. However, the original clip on your hard disk, from which this copy was made, will not be touched.

- If you imported the clip using Leave Files in Place, the alias to the source clip that was imported into the library will be moved to the Trash in the Finder. However, the source clip stored outside the library on your hard disk will not be touched.

FIGURE 14.9 Open the floating Keyword panel by clicking the key icon at the top then add keywords, separated by commas, in the floating Keyword panel.

Clips in the browser can display different color stripes.

- **Red.** The clip was marked as Rejected.
- **Green.** The clip was marked as Favorite.
- **Blue.** The clip has keywords applied to it.
- **Purple.** The clip was analyzed, i.e., for image stabilization.
- **Orange.** The portion of a clip that was used in the Timeline. Enable this display from the View > Browser menu.

Favorites and keywords can also be applied to a range within a clip, which we will cover in a minute as we discuss In and Out. To delete a keyword, select the clip then delete the keyword from the floating Keyword panel.

FIGURE 14.10 To mark a clip, press I to set an In; press O to set an Out. The yellow box defines the range.

MARK To set an In and/or an Out for a clip. If no In or Out is set, the beginning or end of the clip is used.

To find all the clips with a particular keyword applied, click a keyword collection in the Libraries sidebar on the left; for example, Flight. Keywords are a fast and flexible way to organize clips without needing to drag them into separate events.

Step 4: Build the Story

Now that we've imported and organized our clips—which always takes longer than we'd like—it's time to start the fun part: editing. Editing has two parts: reviewing each clip and selecting the part you want to use (this is called "marking the clip," which is shown in **FIGURE 14.10**). The second part is creating a new project and editing clips into it using the Timeline.

If the skimmer is active, it takes precedence over the playhead. I am a big fan of the skimmer in the browser but not in the Timeline—it gets in my way. To toggle the skimmer on or off, press S.

Review and Mark Clips

To review and mark a clip:

- Hover your cursor inside a clip then "skim" left and right.
- Click anywhere in the clip and press spacebar to play the clip. (Press Shift+spacebar to play in reverse.)
- To set an In point (In), which marks the start of a section you like, position the skimmer, or playhead, on the frame you want and press I.
- To set an Out point (Out), which marks the end of the section you like, position the skimmer or playhead on the frame you want and press O.
- To change an In or Out, either retype it or drag a vertical yellow edge.

This process of setting Ins and Outs—which is called "marking"—is repeated for every clip, which is why Final Cut makes marking clips both fast and flexible.

The process of reviewing a clip is easy to write: play the clip, set an In, then set an Out. The process is easy; the hard part is deciding which part of each clip you like and the order you want to build the clips. Final Cut can't help you with the thinking, only the doing.

Selections of clips or other elements are indicated by yellow outlines throughout Final Cut.

Quick Review Shortcuts

To remove the In and Out from a clip, either drag the yellow bars to the edge of a clip or, my preference, press Option+X.

To select the entire clip, put the skimmer or playhead in it and press C.

In general, a clip can have only one In and one Out at a time.

Create a New Project

With clips reviewed and marked, it's time to edit them into...wait! We haven't created our project yet. It's time to do that. Choose File > New Project (Shortcut: Cmd+N) (**FIGURE 14.11**).

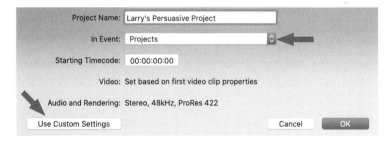

FIGURE 14.11 Choose File > New > Project. Give it a name, pick an event to store it in, and click OK.

There are several ways to create a project, but this is the easiest: File > New > Project. Make sure the button in the lower left says "Use Custom Settings." If it doesn't, click the Use Automatic Settings button. Give the project a name and pick an event to store it in. I always create an event called "Projects" for my projects. It makes them easier to find. Click OK when done.

The Timeline opens to display a new project. (I'd show you a screen shot, but it's another big block of basic black.) In the middle of the Timeline is a darker bar of black. This is the Primary Storyline, and all your clips will head there first when you edit them. It's what makes Final Cut different from other editors. I'll explain what it does in a minute.

You cut from one shot to the next when you want to show the audience something new.

Editing

Editing is the process of moving clips from the browser into the Timeline in a certain order that tells a story. You cut from one shot to the next when you want to show the audience something new. If there's nothing new to show, don't cut.

You can easily drag any clip from the browser to the Timeline, but dragging is slow. A faster way is to click one of the four editing icons, as shown in **FIGURE 14.12**, to edit clips into the Timeline. And, after you've clicked the same icon twice, you may want a keyboard shortcut, which, for me, is the fastest way to edit.

FIGURE 14.12 Buttons at the top left of the Timeline, from left to right: Show/Hide Timeline Index, Connected Clip edit (Q), Insert edit (W), Append edit (E), Overwrite edit (D), Audio/Video edit arrow, and Tool palette (shortcuts vary by tool).

- **Append** edit (Shortcut: E). Edits a clip into the Primary Storyline and places it at the end of all clips in the Timeline. This is the most frequently used editing option.

- **Insert** edit (Shortcut: W). Edits a clip at the position of the playhead (or skimmer, which is why I turn the skimmer off) and pushes any clips after the playhead to the right. This edit always changes the duration of the project.

- **Overwrite** edit (Shortcut: D). Edits a clip at the position of the playhead and deletes whatever clips were under it. This edit generally doesn't change the duration of the project.

- **Connected** clip (Shortcut: Q). Places a clip at the position of the playhead, but on a higher layer. This is the best option when you have someone talking on camera then want to show a picture of what they are talking about; this "picture clip" is often called "B-roll," which is an old film term, and placed on a higher layer.

When you edit clips into the Primary Storyline, each clip acts like a "magnet," attracting any downstream clips so they touch. (Apple calls this the "Magnetic Timeline.") One of the big problems with other editing systems is that it is easy to accidentally leave gaps, which cause flashes of black, between clips. The Magnetic Timeline prevents this.

Talking head. A clip with a close-up of someone talking; for example, a pilot talking about flying. "Talking head" refers to the visual element of a clip.

Sound bite. A clip with audio, generally dialogue, that the audience needs to hear. The pilot's clip is often called a sound bite. "Sound bite" refers to the audio portion of a clip.

B-roll. Media that shows what the speaker is talking about. B-roll would be shots of the plane flying as the pilot talks about it.

Title. Text displayed on the screen. Titles are used to tell the audience something they won't learn from the dialogue, such as the name of the pilot.

When you edit a clip to a higher layer, it connects with the clip below it on the Primary Storyline. This means that if you have someone talking in the Primary Storyline then add B-roll on top, when you move the clip in the Primary Storyline somewhere else in the Timeline, the B-roll clip connected to it moves too.

These two concepts, Magnetic Timeline and connected clips, are unique to Final Cut Pro X. They are designed to solve problems like flashes of black, audio and video clips getting out of sync, and B-roll getting out of sync with the talking head that refers to it. The controversy over this—and there was a *lot* of controversy when FCP X was first released—was that traditional editors weren't used to this behavior. This is one of the reasons I recommend Final Cut for new users. It solves problems automatically that other software requires to be solved manually.

The Rule of Threes

Just like framing a shot, the sequence of shots you use and how you transition between them has meaning, separate from the content of the shots themselves. Filmmakers call this the "grammar" of film.

Film editor Norman Hollyn called the grammar of shot order the Rule of Threes. He said the shot you are looking at now is informed by the shot that came before it and the shot that comes after it. "We don't watch a single clip in a vacuum, we watch a series of clips in context with each other, whose order conveys meaning."[2]

Edit your clips into the Timeline (**FIGURE 14.13**) in the order you want viewers to watch them. My general suggestion is to edit your story in multiple passes; don't try to do everything all at once.

- **Pass 1.** Edit your clips by focusing on what they say, not on how they look. I call this "creating a radio edit." Yes, edit the video with the audio, but concentrate on the story they tell, not specific images. Once the story is complete, go back and clean up the visuals.

[2] (Hollyn, 2008)

- **Pass 2.** Add B-roll to cover edit points and illustrate what the speakers are talking about. This is the phase where we concentrate on the visuals, because the story is reasonably complete.
- **Pass 3.** Add titles and transitions.
- **Pass 4.** Add all remaining effects.

FIGURE 14.13 Four clips edited into the Primary Storyline in the Timeline. To make these easier to see, I also selected them (Cmd+A), which isn't necessary for the editing itself.

The benefit to this system is that you are focusing on one thing at a time, for example, telling the story, illustrating the story, or polishing the story. Otherwise, there is so much to do you get distracted and don't finish anything fully.

The next step, once the basic story is complete, is to add B-roll. B-roll images illustrate what a speaker is talking about. That's where the connected clip icon comes in. It places a clip on a higher layer and "connects" it to a clip on the primary storyline, as shown in **FIGURE 14.14**.

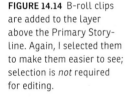

FIGURE 14.14 B-roll clips are added to the layer above the Primary Storyline. Again, I selected them to make them easier to see; selection is *not* required for editing.

In the case of our pilots, we want to hear them talking about flying *and* see them actually flying the planes. Viewers are easily able to hear one thing while looking at something else. This is a common editing technique.

Select each clip you want to use in the browser; then click the Connected Clip icon (Shortcut: Q) to put it on the layer above the talking heads. Do this until you've illustrated their story. As before, drag clips around to get them in the order you like.

Stacking and Visibility

By default, all clips in Final Cut are full-screen. What this means is that whatever clip is on the highest layer will be visible and block the visibility of any clips below it. This is why we put B-roll clips above talking heads. B-roll covers the images of the talking head, without blocking the audio. (All audio is audible all the time. Stacking order has no effect on audio clips.)

Final Cut makes it fast and easy to get clips from the browser into the Timeline. That isn't the hard part of editing. The hard part is figuring out the order of your clips and the best way to use them to tell your story. This often takes lots of trial and error, as well as stepping back and watching your program with fresh eyes. FCP makes the editing process easy, but, again, the thinking is up to you.

Adjust Audio Levels

Part of the editing process is adjusting audio levels so the viewer hears what we want them to hear. This isn't the full audio mix at this point; rather, we are "roughing in" the audio levels so we can hear what's being said. We'll do the full audio mix when the video edit is complete.

To set levels, drag the thin, horizontal yellow line, located in the audio waveform of each clip and shown in **FIGURE 14.15**, up (to make the volume for that clip louder) or down (to make it softer). At this stage, your audio levels don't need to be too precise.

FIGURE 14.15 Adjusting audio levels is part of the editing process. Grab the horizontal yellow line, which represents the audio level of a clip, and drag it up to raise the volume or down to lower it.

The audio meters (**FIGURE 14.16**) are hidden by default, which makes no sense to me. To display them, choose Window > Show in Workspace > Audio Meters (Shortcut: Shift+Cmd+8).

When and Where to Mix

If I'm editing a simple project that's mostly dialogue on a short deadline, I'll do all the audio mixing in Final Cut. There aren't a lot of clips, and FCP has adequate tools for adjusting levels.

However, if I'm doing a complex project with lots of dialogue, sound effects, and music, I'll use Final Cut to edit them into the Timeline so the timing is perfect then move my project into Audition or ProTools for the final mix. Just as Final Cut is optimized for video editing, audio mixing software is optimized to make your audio sound as good as it can.

As we learned in Chapter 12, "Sound Improves the Picture," audio levels don't matter until it is time to export. Still, get in the habit of setting your mix to bounce the meters between −3 and −6 dB. Remember, the more clips playing at once, the louder the audio. Audio must never export with the meters bouncing over 0 dB.

FIGURE 14.16 The audio meters are the *only* accurate judge of levels. You want the total mix bouncing between −3 and −6 dB and never over 0 dB.

Deleting a clip from the Timeline does not delete the clip from the browser, library or hard disk.

Adjust the Timeline Display

To adjust how clips are displayed in the Timeline, click the small clip icon in the top-right corner of the Timeline. The icons determine how big to display waveforms, while the sliders determine Timeline scale and clip size.

Step 5: Organize the Story in the Timeline

At any point in the editing process, you can change the order of clips by dragging them where you want them to appear. Similar to moving children's blocks, shuffle the clips until your story makes the most sense.

If a clip doesn't move when you drag it, make sure you have the Arrow tool (Shortcut: A) selected. This general-purpose tool selects and moves elements. Apple calls this the "Select" tool. This is the most frequently used tool in Final Cut.

In the Timeline, drag clips to move them. This either adds them to an existing layer or creates a new one. Whichever clip is on top is the one you see in the Viewer. (Later in this chapter, you'll learn how to create a "picture-in-picture" where you can see multiple images on-screen at once.) To hide a clip so you can see what's beneath it in the Timeline, press V. (V toggles clip visibility.)

To delete a clip from the Timeline, select it and press the Delete key. If the clip is on the Primary Storyline, all the downstream clips will close up so there's no gap. If any clips are connected to the deleted clip, they will be removed as well.

If, instead of closing the gap, the clip in the Primary Storyline is replaced by a dark gray clip, which displays black in the Viewer, you pressed the wrong Delete key. The big Delete key, above the Return key, deletes the clip and deletes any gap. The small Delete key, next to the End key on full-size keyboards, replaces the clip with a black gap. (If you are on a laptop, press Fn+Delete to emulate the small Delete key.)

FIGURE 14.17 Any clip in the Timeline can be replaced by any clip in the browser.

You can also replace any clip on the Timeline with any clip in the browser (**FIGURE 14.17**).

- Set an In for the browser clip where you want the new clip to start.

- Drag the browser clip on top of the Timeline clip you want to replace.

- In the resulting pop-up menu, select Replace if you want to use the duration of the browser clip. Select Replace from Start if you want to use the duration of the Timeline clip.

The Timeline clip will be replaced by the browser clip, starting at the In of the browser clip.

Copy/Paste also works. By default, when you paste a clip, it will paste at the position of the playhead into the Primary Storyline. To paste to a higher layer, choose Edit > Paste as Connected Clip.

By now you probably noticed that when you move clips around, they snap together but don't overwrite each other. At least, that's the way they behave in the Primary Storyline. If you want to move a clip and have it overwrite another clip, first select the Position tool (Shortcut: P) from the Tools palette (**FIGURE 14.18**). The Position tool essentially turns off the "magnetic" aspect of the Timeline.

FIGURE 14.18 The Final Cut Tools palette.

To cut a clip into multiple chunks, select the Blade tool (Shortcut: B). However, I prefer the much faster Cmd+B, which cuts all selected clips at the position of the playhead. Slicing, dicing, and moving clips is what this phase of editing is all about.

OVERWRITE This means to erase part of an existing clip when a new clip is moved over it on the same layer.

Step 6: Trim the Story

Organizing is about moving entire clips around. Trimming adjusts where two clips touch; the leading and trailing edges of clips to make the transition from one clip to another as smooth as possible. I once spent 45 minutes trimming one edit point before I was satisfied. It was a cut from a wide shot to a close-up of a snowboarder doing a jump; getting that cut right was maddening.

TRIMMING This adjusts where two clips touch, called the "edit point."

We can trim the end of the outgoing clip, the end of the incoming clip, or both the Out and the In at the same time. Trimming makes clips longer or shorter, depending upon how you drag the edit point.

Handles

Handles are essential to all trimming. This is extra footage before an In or after an Out that allow us to adjust the length of clips or add transitions, such as dissolves. When you select the end of a clip, if it turns yellow, you have handles (**FIGURE 14.19**). If it turns red, you don't. If the handle is red, you can make a clip shorter but not longer because you have used all the media at that end of the clip.

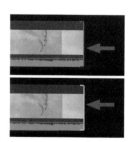

FIGURE 14.19 When the end of a clip is red, there is no more media; it's the beginning or end of the file. When an edge is yellow, the clip has handles.

Another way to see handles, and one that I teach beginning students, is the Precision Editor. Double-click any edit point in the Timeline to open it.

According to Oscar-winning film editor Walter Murch, A.C.E., when it comes to trimming, there are six main criteria for deciding where to cut. He called it his "Rule of Six."

1. **Emotion.** How will this cut affect the audience emotionally at this particular moment in the film?

2. **Story.** Does the edit move the story forward in a meaningful way?

3. **Rhythm.** Is the cut at a point that makes rhythmic sense?

4. **Eye trace.** How does the cut affect the location and movement of the audience's focus in that particular film?

5. **Two-dimensional place of screen.** Is the axis followed properly?

6. **Three-dimensional space.** Is the cut true to established physical and spatial relationships?

Murch made emotion the most important of all of categories in the list.[3]

UPSTREAM This means to the left, upstream, or earlier in the Timeline.

DOWNSTREAM This means to the right, downstream, or later in the Timeline. These are generally used in reference to the position of the playhead or a selected clip.

The Precision Editor, shown in **FIGURE 14.20**, provides the finest illustration of handles I've ever seen. The upper clip is the outgoing clip. Darkened media after the Out are its handles; extra media after the Out. The lower clip is the incoming clip. Darkened media before the In are its handles. You can trim a clip until you run out of handles, meaning there is no more available media. Drag the edge of a clip to trim just that clip. Drag the gray slider between them to trim both the In and the Out at the same time. Press the Return/Enter key to exit the Precision Editor.

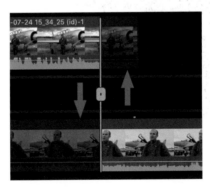

FIGURE 14.20 The Precision Editor. Double-click any edit point to open. Press Return to close.

Trimming

To trim a clip, select an edge with the Select arrow (Shortcut: A) and drag it left or right. If you do this to a clip in the Primary Storyline, all the clips downstream of the playhead move as you drag the edge of a clip.

If you drag the edge of a clip that isn't on the Primary Storyline, the clips next to it don't automatically tighten. Instead, you'll either leave a gap between clips or push other clips out of the way as you move the edge.

[3] (Murch, 2001)

Sometimes we want to trim both the audio and video elements of a clip together. More often, though, we need to trim the video or audio separately. Final Cut makes this easy.

Double-click an audio waveform to separate the audio from the video, as illustrated in **FIGURE 14.21**. Though separate, the two elements are still locked in sync. Double-click the audio waveform to close it back into the video clip.

When the two elements are separate, you can trim the video or the audio, as Figure 14.21 illustrates.

To trim both the In and the Out at the same time, select the Trim tool (Shortcut: T) then drag the edit point to a new location.

While Final Cut provides lots of different ways to trim, dragging the edges are the easiest. If you are like most editors, you'll spend more time trimming than editing. Trimming is that important.

FIGURE 14.21 We can trim audio and video together or separately. Double-click the audio waveform to adjust only the audio.

Other Trimming Tricks

- To delete a clip, select it and press the Delete key.
- To trim an edit point frame by frame, select it then press comma or period.
- To trim an edit point in larger chunks, select it then press Shift+comma or Shift+period.

Step 7: Add Transitions

The first six steps in this editorial workflow I call "creating the story," because they focus on telling a story. It is essential we create the story first, before moving to effects, because without a story, there's nothing to interest the viewer.

The last six steps I call "polishing the story," because now we add the transitions, text, and effects that make everything look nice. We start with transitions.

There are three types of transitions, each of which has an underlying meaning, in addition to changing the shot.

- **Cut.** An instantaneous change in perspective.
- **Dissolve.** A change in time or place.
- **Wipe.** These break the story and take you somewhere else.

You *cut* between two people having a conversation. It occurs at the same time and same place. You *dissolve* between two people talking in the morning, then going out on a date in the evening. You *wipe* between the animated open of a program and the start of the program itself. A wipe says something major is changing, so pay attention.

FIGURE 14.22 To add a dissolve, select an edit point and press Cmd+T. To change its speed, drag a vertical yellow edge.

By default, all transitions are cuts. To add a dissolve, select an edit point and press Cmd+T (**FIGURE 14.22**). To change the speed of the dissolve, drag a vertical edge of the dissolve icon longer or shorter. (The wider the icon, the slower the dissolve.) If you put a dissolve between two clips, there needs to be sufficient handles on each clip for the entire length of the dissolve; otherwise, Final Cut will display an error message.

Wipes are the sexiest, most visually interesting transitions, but they need to be used sparingly. Final Cut has more than 200 to choose from. Click the Transition Browser icon at the top-right corner of the Timeline to display the options (**FIGURE 14.23**). Transitions are grouped by category. When searching, be sure to select the All category first (top-left arrow).

To apply a transition, drag it onto the edge of the clip(s) in the Timeline to which you want to apply it. To change its length, just like a dissolve, drag an edge longer or shorter. To delete a transition, select it and press Delete.

FIGURE 14.23 Transitions browser. Click the icon in the top-right corner of the Timeline to open (top-right arrow). Use the Search box (bottom arrow) to search for a transition by name.

To delete a transition applied to clips in the Timeline, select the transition icon then press Delete.

Step 8: Add Text and Effects

With our story complete and transitions added, we move on to adding text and effects. Adding and perfecting effects will fill more than all the rest of the time you have. There is *always* something you can tweak. This is why we get everything else done first; otherwise, we have a great opening shot and nothing after.

The Inspector, which we are about to meet, is central to all effects. This is where we make changes to selected clips, transitions, and effects. Click the far-right button (red arrow in the figure) at the top right of the Final Cut interface to toggle it open or closed (Shortcut: Cmd+4). The other two buttons hide/show the browser and Timeline.

Double-click the title at the top of the Inspector ("Wing Cmdr. Milne" in my example) to expand the Inspector to full height.

The buttons at the top left of the Inspector vary depending upon what is selected in the browser or Timeline. These include the following:

- Text animation (far left)
- Text formatting
- Generator formatting and animation (not shown)
- Transitions (not shown)
- Video
- Color
- Audio (not shown)
- Metadata (the letter "i" on the right)

Add Text

When it comes to adding text, again, there are hundreds of templates to choose from in 2D, 3D, and 360° formats. (I'll cover 3D text formatting later in this book.) Click the Titles Browser icon in the top-left corner of the interface (shown by the left red arrow in **FIGURE 14.24**). Titles are grouped by category. You can either scroll through the list or search for a title by name (right arrow).

When you find a title, either select it and click the Connected Clip edit button (Shortcut: Q) or drag it into the Timeline. Q is faster, though dragging allows you more flexibility in placement, as the keyboard shortcut always puts clips at the position of the playhead (or skimmer, if enabled).

FIGURE 14.24 The Titles Browser. Titles are grouped by category (left) and searchable by name using the Search box (top right).

FIGURE 14.25 The Text Inspector determines text formatting, as well as providing a window in which to modify your on-screen text.

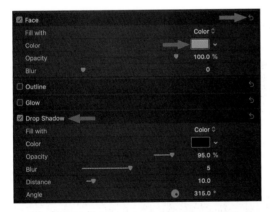

FIGURE 14.26 The text formatting controls at the bottom of the Text Inspector.

FIGURE 14.27 The Text Animation Inspector controls animation specific to each text clip. Not all text has them, but most do.

To change the position of text on the screen, select the text clip in the Timeline, then drag it in the Viewer. Other on-screen controls may be available, depending upon the text clip you chose.

The Inspector provides a wealth of options to make format changes to a title (or a transition, effect, or clip, for that matter). Select the text clip in the Timeline, open the Inspector (top-right blue button), and click the second-from-the-left icon to display the text formatting controls (**FIGURE 14.25**). After using the text controls in Photoshop, these settings should be fairly reassuring; you've used them all before.

Double-click the text in the Viewer to alter the text on-screen. You can modify text in either the Viewer or the Inspector. The Viewer is faster; the Inspector provides more controls.

The Text Inspector also provides a window, at the top, where you can edit the text itself. This is because some text effects create very small text that is hard to edit in the Viewer. This panel provides an alternative way to enter or revise text.

Lower in the Text Inspector are the text format controls. **FIGURE 14.26** illustrates key settings: The "hooked" arrow (top red arrow) resets any setting to its default. Click a color chip to change the color (middle red arrow). Enable an effect by checking the checkbox (bottom red arrow).

My recommended drop shadow settings for text are also illustrated in Figure 14.26.

Almost all transitions, titles, generators, and effects are animated. To determine what animation is available, select a title clip in the Timeline then click the Animation button on the left side of the Inspector (**FIGURE 14.27**). The icon and functions change, depending upon what is selected in either the browser or the Timeline. Text effects vary, too, but Build In and Build Out are common.

Build In and Build Out determine whether to play the opening animation (Build In) or closing animation (Build Out) created for that text clip. These are enabled by default. To disable them, uncheck them. It is easy to see what these controls do by playing the clip. For now, it is sufficient to know where these controls are, how to access them, and how to reset them to their defaults (click the "hooked" arrow on the right side of each effect).

Add Effects

After all our text is complete, it is time to add effects. There are hundreds available in a variety of locations inside Final Cut:

- **Video Inspector.** This manipulates clips, including image stabilization.
- **Effects Browser—Video.** This contains blurs, looks, and stylized effects of all kinds. There are more than 300 to choose from.
- **Effects Browser—Audio.** This contains audio effects that can be applied to individual clips.
- **Clip speed changes.** Located in the Modify > Retime menu, these settings create slo-mo, still frames, and fast-motion.
- **Generators.** Located in the Titles/ Generators menu, these are full-screen animated backgrounds, suitable for info graphics.

I've written thousands of tutorials about these, which are available on my website (LarryJordan. com). In this chapter, I want to showcase key effects available in:

- Video Inspector
- Effects Browser
- Audio Enhancements
- Audio Limiter

Video Inspector Effects

Video Inspector (**FIGURE 14.28**) effects change the size, position, and rotation of a clip. It also provides image stabilization and automatic image scaling, which are also useful.

Adding and perfecting effects always takes all the time you have, plus a bit more. Add effects only when your story is complete.

FIGURE 14.28 The Video Inspector, top-right blue icon, modifies video clips for position, scale, rotation, opacity, blend modes, and more.

FIGURE 14.29 Handheld video should be stabilized. To do so, enable Stabilization in the Video Inspector. Automatic is a good setting to try first.

FIGURE 14.30 Spatial Conform automatically scales (sizes) your visuals to fit the frame. Most of the time, the default setting is the best choice.

To change the size of a frame, select the clip in the Timeline and adjust the Transform > Scale setting. Like Photoshop, digital video images are bitmaps. You can scale them smaller, but scaling video frames larger than 100 percent will cause the image to get blurry.

To change the rotation of an image, move the Transform > Rotation wheel, or enter a value in degrees. To change the position of an image, you can adjust the Transform > Position coordinates, but, in the "Picture-in-Picture" section, I'll show you a much easier way by using controls in the Viewer itself.

If an effect setting is not displayed, roll the cursor over the name of an effect, for example "Stabilization" in Figure 14.29. This reveals the Show button to the right of the parameter name. To hide the details of a setting, click the Hide button. Show/Hide are available only as the cursor rolls over the name of the effect.

Video not shot on a tripod should be stabilized. Enable Stabilization at the bottom of the Video Inspector (FIGURE 14.29). Stabilization will keep the handheld "look" but remove the more distracting bumps and wobbles. Several settings are available, but Automatic is a good first choice.

By default, Spatial Conform (FIGURE 14.30) sizes all images to fit the frame size of your project. But, sometimes we shoot larger frame sizes, like 4K on an iPhone, because we want to vary the image size during editing. In this case, select the clips you want to adjust in the Timeline and change Spatial Conform from Fit (the default) to None. This displays the frame at 100 percent size, regardless of the project settings. At this point, you can adjust the size using Transform > Scale at the top of the Video Inspector.

Picture-in-Picture

A specific effect that's created in the Inspector is picture-in-picture. FIGURE 14.31 illustrates this process. To start, stack two, or more, clips above each other in the Timeline then change the size of the *top* clip. You could do this using the Transform > Position controls. But a much faster and easier method is to select the top clip then click the small arrow in the lower-left corner of the Viewer—shown in Figure 14.31—and select Transform. Blue dots now surround the selected clip.

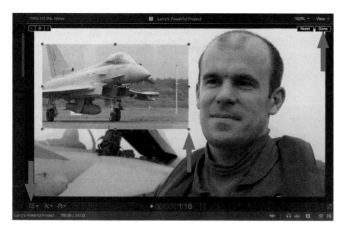

FIGURE 14.31 Picture-in-picture means you place two, or more, images on the screen at the same time, using the Transform settings.

In the Viewer:

- Drag a blue corner dot to resize the frame.
- Drag the middle of an image to reposition it in the frame.
- Drag the blue dot attached to the center circle to rotate the image.
- When the adjustments are done, click Done in the top-right corner of the Viewer.

There's no limit to the number of visuals you can put on screen; the only requirement is that they all be stacked above each other in the Timeline.

Using the on-screen Viewer controls provides a fast, easy way to create the effects you want simply by adjusting the image on-screen. (The settings in the Inspector are more precise, while the Viewer is faster.) Which you use is up to you; I use both all the time.

The Effects Browser

The Effects Browser contains more than 300 effects for video and audio clips, as shown in **FIGURE 14.32**. When you open the browser, a list of effect categories is displayed on the left. Click a category and see all the effects relating to that category on the right.

For example, a really nice effect is Blur > Gaussian. I use this a lot to blur the background so text pops in the foreground. To apply an effect, either:

- Drag the effect on top of a Timeline clip to which you want to add the effect.
- Select all the clips to which you want to apply the effect then double-click the effect in the Effects Browser.

There is no limit to the number of effects you can apply to a single clip.

If an effect's settings are not visible, hover the cursor over the effect name then click the word "Show" to reveal them. Click Hide when you are done to make them disappear.

FIGURE 14.32 The Effects Browser (Shortcut: Cmd+5) contains hundreds of audio and video effects. Open it by clicking the icon at the top right (top-right red arrow). Search for an effect using the Search box at the bottom.

To modify any effect, select the clip that contains the effect, open the Inspector (Cmd+4), click the Video Inspector tab (**FIGURE 14.33**), find the effect, and tweak. The cool thing about effects is the "right" setting is the one that looks good to you.

The Inspector provides a variety of settings for each effect:

- To reset an effect, click the small arrow to the right of the name and choose Reset Parameter (**FIGURE 14.34**).
- To hide the effect settings, click Hide.
- To disable an effect, without removing it or resetting its settings, uncheck the blue checkbox.
- To delete an effect, select the effect name in the Inspector and press Delete.

FIGURE 14.33 To modify any effect, select the clip in the Timeline, open the Inspector (right arrow), click the Video tab (left arrow), find the effect in the stack, and tweak.

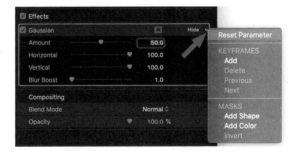

FIGURE 14.34 To reset all effect settings to their default values, click the small arrow to the right of the effect name and choose Reset Parameter.

Effect Stacking Order Makes a Difference

Effects in the Inspector process from top to bottom. Changing the "stacking order" will change the effect, as illustrated here. In the image on the left, first we add a blue border, then we remove all the color. This creates an overall black-and-white image.

In the image on the right, first the color is removed, then the blue border is added. The settings for both effects—and the original image—are the same. All that changed was the order of the effects applied in the Inspector.

If you don't get the effect you were expecting, change the stacking order—by dragging the name of the effect up or down—and see whether that makes a difference.

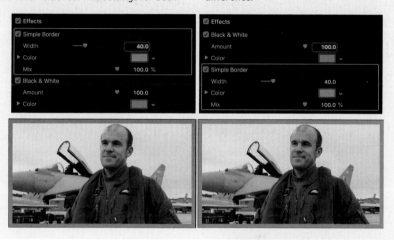

Audio Enhancements

While Final Cut is not as good as Adobe Audition in mixing audio projects, it does have a lot of helpful audio tools, especially when you are in a hurry. Several of these are grouped under Audio Enhancements that allow you to adjust levels, remove hum, EQ a clip, and reduce noise.

To apply these, select the clip containing audio and choose Modify > Auto Enhance Audio (Shortcut: Option+Cmd+A). As **FIGURE 14.35** illustrates, you can inspect your clips in the Audio Inspector for problems. If no problem is detected, Audio Enhancements shows a green checkbox. If there are problems, Audio Enhancements can help fix EQ, Loudness, Noise Removal (in spite of its name, this *reduces* noise; it won't remove it), and Hum Removal (hum can be entirely removed). To enable or disable any setting, check the blue checkbox.

FIGURE 14.35 The Audio Enhancements options in the Audio Inspector (top red arrow). The top image shows a clip with no audio problems; the bottom image shows the adjustments available to fix poor audio.

Audio Limiter

In Chapter 12, "Sound Improves the Picture," I illustrated the Multiband Compressor as a way to boost and smooth audio levels. Final Cut has something similar but simpler to use: the Limiter effect. This boosts softer dialogue without distorting louder passages. It should generally be applied to dialogue, not sound effects or music.

Ignore the Limiter filter at the top of the Levels category. It lacks a key setting that makes this version of the filter less useful.

To apply the effect, go to the Effects Browser (Shortcut: Cmd+5), as shown in **FIGURE 14.36**. Scroll down until you see the Audio section and, within that, the Levels category. Scroll through the effects on the right (which are sometimes called "filters") until you see the subcategory Logic. Inside that, look for Limiter. (Whew! That was a lot of scrolling.) Finally, as with all effects, drag it from the Effects Browser and drop it on the clip you want to adjust.

To adjust the Limiter, select the clip that contains the effect in the Timeline, go to the Audio Inspector (see **FIGURE 14.37**), and click the small icon on the right side of the Limiter effect.

FIGURE 14.36 Audio effects are stored in the Effects Browser > Audio section.

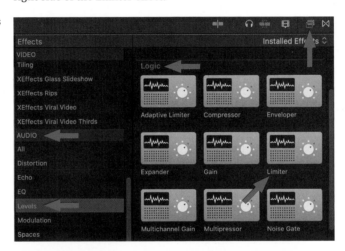

FIGURE 14.37 Open the Audio Inspector (top arrow), make sure the Limiter effect is enabled (lower-left arrow), then click the Limiter controls icon (right arrow).

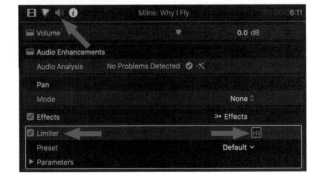

Similar to the Multiband Compressor in Audition, the Limiter in Final Cut raises the level of softer passages in a clip, on an instant-by-instant basis, until the volume reaches the Output Level value. This is the upper "limit" of the audio level, from which the effect gets its name.

FIGURE 14.38 illustrates the Limiter effect. Four settings in this effect always use the same values:

- Set Release to 500 ms.
- Set Output Level to −4 dB.
- LookAhead remains at 2.0 ms.
- Set True Peak Detection to On.

FIGURE 14.38 This looks more imposing than it is. Four settings always use the same values; you only need to adjust Gain for each clip.

The only setting you need to adjust is Gain. With these settings, your audio will never get louder than −4 dB for this clip. To adjust the Gain setting, play the clip and watch the Reduction column. Increase the Gain until you see the Reduction level bouncing between −1.5 and −3 in the loudest passages (Figure 14.38). (A little more won't hurt; a lot more will sound blatty and unpleasant.) There may be times during playback where the Reduction level doesn't bounce at all; as long as it bounces at some point during the clip, your settings are fine. What you want to do is make sure all your levels are increased without distorting the loudest passages.

And that's it. The Limiter filter keeps an eagle-eye on your levels, without you having to spend a lot of time adjusting each clip, just as the Multiband Compressor does in Audition.

Step 9: Create the Final Audio Mix

We talked about audio a lot in Chapter 12, and I won't repeat that information here. Your goal, during the audio phase, is to create the audio environment the audience will hear.

- Design the sonic environment, also called "sound design."
- Edit existing audio so that dialogue is clean.
- Add sound effects and music.
- Mix all the different audio elements and make sure that levels bounce between −3 and −6 dB and that they never exceed 0 dB.

To restate what I mentioned at the beginning, mix simple projects directly in Final Cut and export complex mixes to audio mixing software. Software to consider includes Avid ProTools, Adobe Audition, Apple Logic Pro X, Cockos Reaper, and PreSonus Studio One.

Step 10: Finalize the Look and Colors

There are two ways to work with color: to fix problems, called "color correction," and to achieve a look, called "color grading." Let's start with color correction.

There is a wealth of manual color correction tools available in Final Cut. Peachpit Press has several books that show how to use all these different color tools.

Simple Color Correction

You never plan to have problems with your colors, but, when you do, there's a fast technique to fix them.

FIGURE 14.39 shows a typical color problem: a clip that has a color cast—in this case, it's too green. To fix this, select the clip with the color problem then choose Modify > Balance Color (Shortcut: Option+Cmd+B).

FIGURE 14.39 A typical color cast: a clip that's too green.

In the Video Inspector (**FIGURE 14.40**), change Method to White Balance then click something in the image that's *supposed* to be white or medium gray. In this example, I clicked the underside of the tail. This effect figures out how to change the colors in a clip to get the colors right.

The results of using this tool are stunning. **FIGURE 14.41** compares the before (left) to after (right). While you can spend a lot of time in Final Cut creating captivating looks, sometimes all you need to do is fix a problem. The Balance Color tool does exactly that.

FIGURE 14.40 In the Video Inspector, change Method from Automatic to White Balance.

FIGURE 14.41 The corrected result, on the right, is stunning. And instant.

Color Grading – Create a Look

In addition to fixing color problems, there are more than four dozen color presets (Effects Browser > Color Presets, Comic Looks, and Looks) that can create dramatic color treatments for your clips (**FIGURE 14.42**). You apply them the same as any other effect: drag one you like from the Effects Browser and drop it on a Timeline clip.

FIGURE 14.42 Here are 12 of more than 25 looks to create a color treatment that best represents your message to the audience.

Some of these effects have settings you can tweak, others don't. If you like one, use it. If not, adjust a setting for that look in the Video Inspector to see if that helps. Otherwise, delete the effect in the Video Inspector and try something different. These are fun to help you discover how changing the look also changes your emotional response to it.

There are numerous other color tools in Final Cut that manipulate color in very sophisticated ways. Looks are a simple way to start exploring what's possible.

Step 11: Output the Project

Once all your review, editing, trimming, transitions, titles, and effects are done, it's time to output your project and share it with the world. Apple calls this process "sharing." (I call it "exporting.") Select your project in the Timeline then choose File > Share > Master File.

The first screen that appears, shown in **FIGURE 14.43,** is the Info screen, where you can change the labels associated with the clip or hover your mouse over the image and skim your project to make sure you are exporting the right version.

FIGURE 14.43 The Info screen displays metadata (labels) that you can change. You can also skim the image to make sure you are exporting the right project.

When exporting, the Settings screen, shown in **FIGURE 14.44**, determines what you are exporting. My general recommendation is to always export a master file in the video format that matches your project, in other words, the one at the top of the Video Codec menu with "Source" in the title. Specifically, I suggest using Apple ProRes 422.

Notice in the lower-right corner Final Cut estimates how big the exported file will be. Media files are really big, so make sure you have the room to store it.

FIGURE 14.44 The Settings screen determines the technical export settings.

Now What Do I Do with It?

You've got this master file, so now what do you do with it? The easiest answer is use Apple Compressor to create versions for YouTube, FaceBook, Vimeo, Instagram, local websites, broadcast, cable, digital satellite, and mobile devices of all sorts.

In other words, you will probably need to make lots of versions of this one project. Compressor is great for that. It has presets that simplify the process, and it integrates well with Final Cut. Best of all, it doesn't cost much. You can also use Hand-Brake or Adobe Media Encoder.

And if you need help using it, my website has plenty of tutorials (LarryJordan.com).

Step 12: Archive the Project

The last step is to archive your work. Make sure you have backups of the final master file and your projects. How long you decide to store them is up to you, but don't be in a hurry to get rid of anything.

The general rule, which I've learned the hard way, is that no one wants any file stored on your system until the day *after* you trashed it.

KEY POINTS

The script creates the idea. Production records the story. But we tell the story through editing. This 12-step editorial workflow is an effective way to stay focused and on track. Video editing is part craft, part technology, and part logistics, with a lot of client management thrown in. Here are key thoughts from this chapter.

- Time spent planning the edit and organizing media before you start is never wasted.
- You will always need more storage. No project ever got smaller during production.
- Shoot and edit the frame rate you need to deliver.
- If people can't hear, they won't watch.
- Backups of all project and media files are essential. Editing without backups redefines the term "pain and suffering."
- Keyboard shortcuts will not make you a better editor, but they will make you a *faster* editor.
- Build the story first then worry about effects. (The only reason for fast, jarring edits and distracting effects is to hide the fact that your talent has no talent.)
- Don't add effects or color correction until your story edit is complete. Your audience is not there to watch your effects; they are there to watch your story.
- Always export a high-quality master file to archive, then compress later.
- No video project is ever perfect. Don't strive for perfection; strive to get it *done*.

It takes a lot of work to make something look effortless. And persuasion, to be effective, should not look forced. The best, and most depressing, compliment an editor can receive is, "I really loved your program! It didn't even need any editing."

PRACTICE PERSUASION

Think about editing a conversation where: Each shot is a wide shot or a closeup, or where actors talk over each other or insert long pauses, or where everyone moves or no one moves. How does all this different pacing make you feel?

A LIFE-CHANGING CONVERSATION

Larry Jordan

It was 2003, and I was out of work, again. I saw an ad in the *Hollywood Reporter* that said, "Assistant Film Editor Needed. Must know [Apple] Final Cut." Well, I said to myself, I know Final Cut. Apple just certified me as an Apple Final Cut trainer. So, I answered the ad.

A woman with a French accent answered, and we made arrangements to meet. At the appointed time, I walked up to a small single-story Hollywood bungalow and knocked. A woman in her mid-50s answered and invited me in.

As we walked into the living room, I looked around, admiring the photos on the wall, the comfortable furniture, and the bric-a-brac on the bookshelves.

Françoise, for such was her name, was pleasant, with graying hair, a French accent, and a direct yet pleasant manner. As we chatted, I looked again at the bookshelves and saw several awards, including a BAFTA and an Oscar!

Stunned, I pointed to them and asked, "Wow! An Oscar. What did you get it for?" She smiled and said, "Film editing." Now, you may think, world-famous raconteur and video expert as I am, that I meet Oscar-winning editors all the time. You would be wrong. She was my first.

More importantly, it made me realize that I was in the wrong place. I turned back to her and said, "Françoise, I am not the right person for you. I know Final Cut, and I know video, but I've never worked on a feature film in my life. There's no benefit for you to hire me. However, I would be really grateful if we could spend 15 minutes chatting. At which point, I promise to leave and not bother you again." She very graciously agreed.

Now, it would help my story significantly if I could remember even one word of our conversation during those 15 minutes. I don't get to talk to Oscar-winners often. But I can't, except for one comment that changed my life when I asked, "Françoise, how do you know when your films are done?"

She looked down at her lap then, looking up with a small smile, she replied, "Done? Done? My films? My films are never done. They rip them from my grasp, even though I can still make them better."

In that one answer, I learned there is never enough time for *any* creative project. A work of art is never completed; it is reluctantly abandoned as time runs out.

Her name was Françoise Bonnot, and while we never did work together, her advice changed my life. She edited into her 60s and died in 2018.

How well we communicate is determined not by how well we say things, but how well we are understood.[1]

ANDREW GROVE
CEO, INTEL

CHAPTER 15

MOTION GRAPHICS: MAKE THINGS MOVE

Most of us are fascinated by motion graphics but have no idea how to create them. Motion graphics are everywhere. They are designed to capture the viewer's eye and convey a message.

We're going to look at creating motion graphics (small "m") using software from Apple called Motion (big "M"). The capitalization is important.

My goals for this chapter are to:

- Explain what Motion is and how it is used
- Explore the interface
- Format and animate text
- Create a simple project filled with moving objects
- Create composite images
- Show how to build a basic motion graphic video

[1] Andrew Grove, CEO, Intel quoted in Jennifer Lee, *The Book of Quotes*, Essentially Organized, 2011

RENDER To calculate new media from existing media.

APPLE MOTION CREATES motion graphic videos. So does Adobe After Effects. Why, then, am I discussing Apple Motion? Motion is much easier to learn. After Effects is designed for visual effects professionals. Motion is designed for any creative artist who needs to create motion graphics but can't devote full-time to learning the software.

Our goal is persuasion, and our tactics are to attract the eye of the viewer long enough to deliver a message. If you are trying to reach a distracted audience, nothing grabs them like movement. Remember the Six Priorities in Chapter 2, "Persuasive Visuals"? The number one "eye-catcher" is movement. That's why motion graphics are so popular—all they are *is* movement.

One of the exciting aspects about Motion is that it allows you to create compelling, high-quality videos all by yourself. Unlike video production, no collaboration is needed. Still, creating motion graphics can be intimidating, so take the time to read and practice.

Motion creates videos that can be posted to the web, imported into any video editing application, or displayed on monitors in malls or campuses. In other words, Motion videos can go anywhere.

Another big benefit to using Motion is that Apple designed it for continual, real-time playback without waiting for rendering. To do this, it aggressively pushes the graphics processing unit (GPU) inside your computer. This means, as you build projects, you can watch what you are creating play back in real time. Motion actively encourages you to "try something and see what happens." If you like it, keep it. If it is too ugly for words, undo it, and no one will know.

Why Learn Motion?

A class challenged me one year to explain why they should learn Motion instead of After Effects. Both are excellent software, though After Effects is better known. Here was my response detailing the benefits of Motion:

- One-time, low-cost purchase (After Effects requires monthly fees)
- Optimized for social media
- A quicker learning curve
- Tight GPU integration means effects play in real time without rendering

- An integrated library of more than 1,900 graphic elements
- Very powerful, flexible, and fast text animation, including styled 3D text
- Extensive painting and path tools
- Very flexible particle engine
- Accessible cameras, lights, particles, movement and sets in 3D space
- Extensive replicators, though I don't cover them in this book
- Tight integration with Final Cut Pro X
- Easily create compelling eye candy quickly

One other thought before we start: Motion is unlike any other software you've ever used. It can be frustrating to learn, as I discovered when I was learning it myself. Don't panic! Of all the software I've taught my students, the one they enjoyed learning the most was Motion.

Creating Something Simple

The easiest way to think about Motion is to imagine a new version of Photo-shop where every element can move. Just as with Photoshop, Motion has elements, layers, selections, and filters. And, just like Photoshop, we stack elements to create composite images. So, you already understand the basic conceptual framework of how Motion works—even before you start using the program.

When you start Motion, the Project Browser appears, as shown in **FIGURE 15.1**. Motion plays a dual role: Not only is it a stand-alone motion graphics pro-gram, but it is the front end to creating visual effects templates for Apple Final Cut Pro X. While we won't be covering that aspect of the program in this book, it is worth knowing that Motion can create custom effects for Final Cut.

Motion always creates video at the highest possible quality.

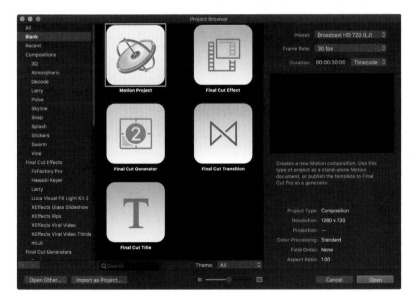

FIGURE 15.1 When you first start Motion, the Project Browser appears.

Creating a New Project

To get started, select Motion Project, configure the settings in the top right, then click Open. We will follow this same process for every example in this book.

FIGURE 15.2 Project settings. Before creating a project, set the duration (left), frame size (Preset menu in the middle), and frame rate (menu on the right).

Motion always creates video at the highest possible quality. The project settings, located in the top-right corner of the Project Browser and shown in **FIGURE 15.2**, determine the frame size, frame rate, and duration of a project. My recommendations, especially for learning the software, are:

- Broadcast HD 1080
- 30 fps
- A Duration of 10 seconds, expressed in timecode as 00:00:10:00

This creates a video with a frame size of 1920 x 1080 pixels, which is compatible with YouTube and Facebook, though not for the square aspect ratio of Instagram.

Once you have the settings you want, click Open.

The Motion interface appears; it's dark, empty, and inscrutable, as you can see in **FIGURE 15.3**. Motion may be the greatest thing since sliced bread, but it sure doesn't look very inviting. Not to worry, we'll get this looking good shortly.

FIGURE 15.3 This is the Motion interface—dark, empty, and inscrutable. Sigh...It's not a great way to make a first impression.

Frame Rates Don't Matter

When you are creating media for the web, the frame rate doesn't matter. (It does, though, if you are creating motion graphics for broadcast, cable, or digital cinema.) The reason I picked 30 frames per second (fps) is that it looks good, creates smaller files than faster frame rates, minimizes motion blur, and simplifies working with timecode. You can pick any frame rate you want, keeping in mind that faster frame rates create larger files. If you plan to integrate a motion graphic video with a normal video project, the frame rates should match.

The Motion interface has four regions, as shown in **FIGURE 15.4**:

1. **Library/Inspector.** This stores a library of visual elements, plus the controls to make changes to them.
2. **Layers panel.** Like the Layers panel in Photoshop, this adds, modifies, reorganizes, and removes elements.
3. **Viewer.** This is where we watch our videos.
4. **Timeline.** This is where we can adjust the timing of elements; however, there's an easier way called the Mini-Timeline, which I'll explain as we go.

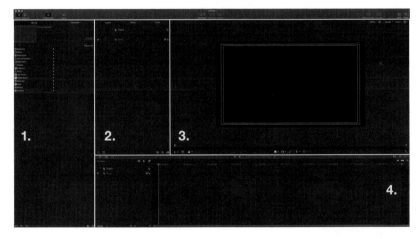

FIGURE 15.4 The four regions of the Motion interface: 1. Library/Inspector, 2. Layers panel, 3. Viewer, and 4. Timeline.

For this chapter, though, we don't need to use the Timeline. So, let's hide it. Go to the Window menu at the top of the interface and uncheck Video Timeline. The lower black block disappears. Everything we need to do, we can do with the remaining three boxes. (Plus, hiding the timeline gives us more room to work.)

FIGURE 15.5 The Text tool is located at the bottom center of the Viewer.

Adding Text

Let's begin exploring Motion by doing something we learned to do in Photoshop: add text. Then, unlike in Photoshop, we'll animate it.

Just as with Photoshop, we select the Text tool to add text (**FIGURE 15.5**). The Text tool (Shortcut: T) is located at the bottom center of the Viewer. (We use this to create both 2D and 3D text, though I'll save creating 3D text until later.) Click anywhere inside the Viewer and start typing.

When you do, two things happen at once: A new layer is added to the Layers panel, and your text appears in the Viewer (**FIGURE 15.6**). When you are done entering text, press Escape (ESC) to exit text entry mode. A box, with blue dots, appears around your text.

FIGURE 15.6 Motion is a very friendly program!

Layers are a key part of Motion. A layer contains one and only one element. Each layer can have its own duration, timing, effects, and filters. We can manipulate the element stored on a layer, the layer itself, or the group that contains the layer. For example, we can change the color of an element, animate the layer, or hide all the layers contained in a group.

We view the timing of elements on the Mini-Timeline. We can also view projects using the main Timeline, which we hid earlier. But the Timeline is an advanced feature that often proves more confusing than helpful in learning the software.

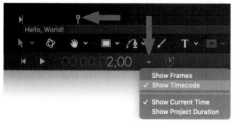

FIGURE 15.7 The blue bar of a video clip is displayed in the Mini-Timeline. The top arrow indicates the playhead. The numbers represent the position of the playhead. Click the down arrow to display the playhead position in frames or timecode.

The Playhead and the Mini-Timeline

When you play a project, the "playhead" (indicated by the top red arrow in **FIGURE 15.7**) moves along the Mini-Timeline, which is located at the bottom of the Viewer. When you select a group, layer, element, behavior, or filter in the Layers panel, its timing is displayed as a bar in the Mini-Timeline. The left edge of the Mini-Timeline represents the start of a project, while the right edge represents the end. Just like Final Cut, to play a project, press the spacebar. To stop, press the spacebar again.

The playhead indicates the current frame we are viewing in the Viewer. Drag the playhead across the Mini-Timeline to skip through your project at high speed. By default, during playback, Motion loops from the end to the beginning of a project and continues playing. Press Home to jump the playhead to the beginning of a project; End jumps it to the end.

The specific location of the playhead is indicated by the numeric display at the lower center of the Viewer. (Video folks tend to prefer timecode, while animators prefer frames. Figure 15.7 shows how to pick your preference.)

Click anywhere just above the blue, purple, or green bar to jump to that part of the Timeline. I'll explain what the clip colors mean at the end of this exercise.

The Mini-Timeline is where we adjust the timing of the selected element.

The Tools Menu

Pressing Escape does more than just exit a text or drawing mode; it also selects the Arrow tool. (Apple calls this the Transform tool; the rest of us call it the Arrow tool.) It's located immediately to the left of the globe thingy. This general-purpose tool selects and moves elements. If you ever find the cursor not doing what you expect, it's probably because the Arrow tool is not active.

To select the Arrow tool, either click it at the bottom center of the Viewer or press Shift+S. The Transform tool is the most used tool in Motion.

Select the text layer for the text you just typed. Then, click inside the bounding box displayed in the Viewer and drag the text where you want to place it. While you can change the size of the text by dragging a blue dot on the bounding box, I recommend instead that you use the Inspector. Why? Because when you scale the text by dragging, you don't really know what size it is, so it is hard to create other text to match the size without spending time dragging. It is better to size the text using the Inspector, which we'll cover next.

Press the Shift key when dragging a blue dot to retain the aspect ratio of the text.

Select the text layer in the Layers panel then, on the top left of the Motion interface, click the word "Inspector." (These words are called "text buttons.") The Inspector is where we make changes to whatever we first selected.

FIGURE 15.8 The text buttons for the Inspector: Properties, Behaviors, Filters, and Text are located under the Preview window. The far-right button (Text) changes depending upon what's selected in the Layers panel.

FIGURE 15.9 The Text > Format menu is similar to the text controls in Photoshop.

Rather than dragging the blue dots to size the text, it is better and potentially higher quality to use the Inspector, or HUD, to alter the text. (We'll meet the HUD shortly.) The Inspector has four text buttons below the preview image, as shown in **FIGURE 15.8**:

- **Properties.** This changes the size, position, and rotation of the selected layer.
- **Behaviors.** This modifies behaviors, which are prebuilt animations applied to the selected layer.
- **Filters.** This modifies filters, which affect the look applied to the selected layer.
- **Text.** This option varies depending upon what is selected in the Layers panel.

Click the Text button and three more submenus show up:

- **Format.** This changes fonts, spacing, and alignment.
- **Appearance.** This changes colors and drop shadows.
- **Layout.** This changes how the text is displayed and is a menu that I almost never use.

The Text > Format menu is, again, similar to Photoshop because both use similar text formatting tools.

Again, with Photoshop in our background, most of the options shown in **FIGURE 15.9** will feel familiar such as Font, Size, Alignment, and Line Spacing. You already know how these work. Modify a setting and see how it changes your selected text.

Animating the Text

Now, let's move into the "magic" part—animating the text. The easiest way to animate anything is using behaviors. (Behaviors are what make creating motion graphics in Motion both fast and fun.) Again, select the layer that contains the text; remember the rule: "Select something, then do something to it."

Change a Position Preference

If you are adding elements to the Layers panel and they start at the position of the playhead in the Mini-Timeline, rather than the start of the project, you need to change a preference. Go to Motion > Preferences > Project and select "Start of project" as shown. I find it easier to move a clip after I create it than to keep remembering to put my playhead in the right spot before adding the element.

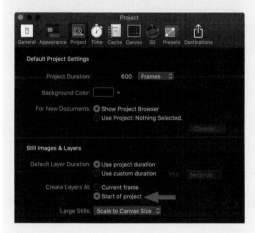

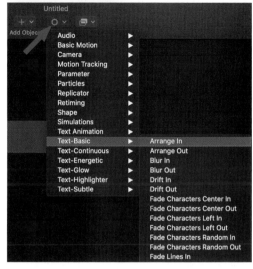

FIGURE 15.10 At the top center of the Motion interface are three buttons: Add Objects, Behaviors, and Filters. Click the Behaviors button to display multiple categories (left) and hundreds of animations (right).

There are almost 200 behaviors bundled with Motion. (You can see some of them in **FIGURE 15.10**.) To access them, click the Behaviors gear icon at the top center of the interface. Behaviors are grouped into categories, with the last seven devoted to text animations.

As you roll over a category, individual animation options appear on the right. If it includes the word "In," the text makes an entrance. If it includes the word "Out," the text makes an exit. It is impossible to describe in words what these do. Select one, play your project, and watch.

BEHAVIOR A small, pre-built animation that can be applied to a group or element that causes it to move without using keyframes or programming.

No Need to Render

Motion aggressively pushes the GPU so that projects play in real time without rendering. This means you can make a change then instantly watch to see whether you like it. Many creators simply let the project play continually as they make changes. There's no need to wait for a render.

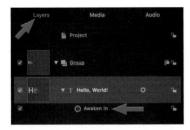

FIGURE 15.11 Each time you add a behavior, it appears indented under the layer to which it is applied.

Each time you add a behavior to anything, it appears as its own layer inset under the layer to which it is applied, as you can see in **FIGURE 15.11**. To disable a behavior, uncheck it. To delete a Behavior, select it and press the big Delete key (the one above the Return key). You can apply more than one behavior to the same layer.

Some of these animations are unbelievably tacky. Others are really nice, and most will make you giggle. Go ahead, play.

I'll wait.

• • •

Interim Summary

Before we go on, let me summarize what we've covered so far:

- You know how to create and configure a new project.
- You know how to add and modify text using the Inspector.
- You know how to find and apply behaviors to animate an element.
- You know how to start and stop playback of a project.

These are key techniques that we will return to again and again.

Creating a Simple Animated Composite

Let's use Motion to create a simple motion graphic composite, say to post to a Facebook page or to display on a monitor in the mall. There's no audio with this (yet), but we will animate the elements (**FIGURE 15.12**).

FIGURE 15.12 This is the composite we will create. Every element is animated, while the Layers panel shows the different components and their effects.

Taking Time to Plan

Before I open Motion and start a new project, I've found it's useful to think clearly about what I want to create. It's that "planning" thing again. Specifically:

- What message do you want the video to convey?
- What is the timing of the elements?

A storyboard helps answer both these questions. As we discussed in Chapter 7, "Persuasive Photos," a storyboard is a sketch of what you'd like the finished project to look like, along with the text you want to use. It can also help determine timings for scenes or the specific elements that appear during the video. Just as in a story, you can't say everything at the beginning. You need to reveal elements over time. A storyboard will help. (I'll cover how to time scenes in the next chapter.)

In a motion graphic video, every second has value. Plan your elements and timing carefully.

Most motion graphic videos are very short—10 to 20 seconds. This means that every second has value. A good way to plan a motion graphic is to divide it into three scenes (maybe four if you have limited text) then give each scene a specific amount of time.

- Scene 1: Establish the problem.
- Scene 2: Develop the problem.
- Scene 3: Provide a solution to the problem with a Call to Action.

For a 15-second video, this means each scene runs only five seconds. Remember in Chapter 3, "Persuasive Writing," when I stressed that writing for video is more like poetry than prose? This timing issue is why. If you have only five seconds per scene, that means your viewers can only read and comprehend five to ten words per scene. Writing paragraphs means they won't get your message.

This is a hard concept for nonvideo people to grasp. My students keep wanting to put a full paragraph of text on the screen or dawdle on the first scene. Too much text means there isn't enough time to read it. Too much time spent on the first scene means that you won't have enough time when you most need it: displaying the Call to Action.

Allow Time for Viewers to Read Your Text

You *must* give your audience enough time to read your text. You read your text each time you preview the project during editing. Your audience, though, gets only one chance to read it. In general, hold text on-screen long enough for you to read it out loud *twice*. If you can't get it read in time, hold it on-screen longer or remove some words.

Put the most arresting image first then end with the Call to Action.

A rule of thumb is to put your most arresting visual first. In this example, it's the title text—big, bright, and with movement to catch the viewer's eye. Then, end with the Call to Action telling the audience where to go and what to do.

Motion Graphics Love Video

Motion graphics are a perfect place to add video, which we will work with in the next exercise. While audio is always important, many motion graphic videos are often watched silently, so make sure your image carries the full story.

GROUPS are used to organize projects and to allow one effect to control multiple elements.

Organizing a Project Using Groups

To start, create a new Motion project. Like Photoshop, composites in Motion often have lots of layers. So, we use groups (which are simply folders to which Apple gave a new name) to organize our projects.

To create a group, right-click in the dark gray area at the bottom of the Layers panel (or choose Object > New Group), as shown in **FIGURE 15.13**. In this example, we'll create three new groups: Text, Shapes, and Background. To rename a group, select it, then press the Return key.

Just as with Photoshop, the stacking order of groups and elements determines foreground to background order, with foreground at the top of the Layers panel.

- To change the stacking order, drag the name of a group, element, or effect up or down.
- To put a group inside another group, drag them on top of each other.
- To move an element into a different group, drag it into the new group.
- To delete a group, highlight it and press the *big* Delete key.

FIGURE 15.13 To create a group, which is another name for a folder, right-click in the dark gray area of the Layers panel and choose New Group.

I keep mentioning the big Delete key, located above the Return key, because the small Delete key, next to the End key on full-size keyboards, is programmed differently. Always use the big Delete key.

Apple has not made moving groups easy. For example, there is no keyboard shortcut to move a group up or down. Instead, you need to drag it, as **FIGURE 15.14** illustrates.

When moving groups, a blue line, illustrated in Figure 15.14, appears indicating the group you are moving. If the blue line is aligned with the left edge of a group (left), it means the group you are moving will go between groups. If the line is indented (center), it means the group you are moving will go inside the group above it. If a group itself is highlighted, it means the group you are moving will go inside the highlighted group.

FIGURE 15.14 The process of moving a group requires closely watching the position of a very thin blue line.

Another helpful technique in designing a motion graphic is to start at the lowest layer of your project and work up. So, the first thing we'll create is the background.

Creating the Background

Click the Library text button in the top-left corner of the Motion interface, then click Content > Backgrounds. Scroll until you find Golden Reflection and click it once (**FIGURE 15.15**).

If you see a list, rather than thumbnails, click the "4 squares" icon at the bottom-right corner of the Library.

This loads it into the small preview window at the top. Select the Background group (the one on the bottom of the Layers panel). Click the Apply button to the right of the preview window, and your background is added to the group and displayed in the viewer.

To make sure you see the entire project image, click inside the Viewer to make it active then press Shift+Z. This resizes the image to fit in the Viewer.

The problem is that many of Apple's backgrounds are too bright. *Way* too bright. So, let's use Levels to dim it back. (You could also reduce the Properties > Opacity setting, but that actually makes the background translucent, which you may not want.)

Filters change the look of an element or a group containing elements. Select the Background group (which means the filter we are about to apply will affect all

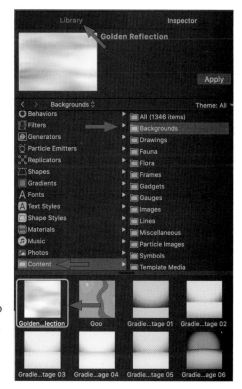

FIGURE 15.15 A world of animated elements is stored in the Library. Backgrounds are in Content > Backgrounds.

There is no limit to the number of behaviors or filters that can be applied to a single element or group.

the elements contained in that group) then choose Filter > Color > Levels (**FIGURE 15.16**). One nice thing, which makes using Levels easier, is that Levels in Motion work the same as Levels in Photoshop.

Select the Levels filter in the Layers panel then choose Inspector > Filters > Levels to adjust the settings. The settings I used are illustrated in **FIGURE 15.17**.

Here's another important point: Behaviors and filters can be applied to an element *or* group. Applying an effect to a group means that one effect can affect *all* the elements inside that group. This is a huge time-saver when lots of elements all need to move together or look the same.

That's it for the background layer. Twirl up the small triangle next to Background in the Layers panel to hide the contents.

FIGURE 15.16 Click Filters to display the different options to change the look of an element. (I've installed some custom plug-ins; your list won't look the same as mine.)

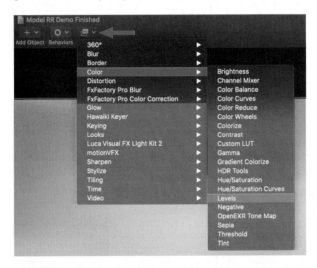

FIGURE 15.17 The Levels filter applied to a group, rather than a single element.

The blue rectangles around the image are called "Safe Zones." The outer rectangle is called "Action Safe," which is 5 percent in from all edges. The inner rectangle is called "Title Safe," which is 10 percent in from all edges. These indicators were adopted decades ago by the video industry to protect essential text, such as phone numbers, and graphics, such as corporate logos, from getting cut off when images were broadcast. We use these today because many video display technologies crop images after they leave your control.

Whenever you create a video for the web, keep all text, URLs, graphics, and other essential elements inside the *Action* Safe zone, which is the outer rectangle. When creating graphics for broadcast, cable, or digital cinema, keep all essential elements inside the *Title* Safe zone. Protect yourself; it's important.

Creating the Mid-Ground

Twirl down the Shapes group in the Layers panel. Yup, it's still empty.

There are three geometric drawing tools in Motion: Rectangles, Circles, and Lines (**FIGURE 15.18**). (We can also draw paths, which I cover in a supplemental online chapter.) They all work in a similar fashion: Select the Rectangle tool and draw a rectangle in the lower third of the frame. The easiest way to draw is to start outside the frame and drag through to create the shape you want.

FIGURE 15.18 The drawing tools are located at the bottom center of the Viewer.

When you are done drawing, press the Escape (ESC) key. This exits drawing mode, just as pressing Escape exits the Text mode.

Click the Arrow tool then select the rectangle in the Layers panel. The rectangle is highlighted in the Layers panel and displays a box around it in the Viewer. Choose Inspector > Shape and change the color to something darkish. I chose a pea green. Next, choose Inspector > Properties and change Blend Mode to Screen. **FIGURE 15.19** illustrates both these settings.

Press Shift to constrain the drawing to a perfect square or circle.

Press Option to draw the object from the center.

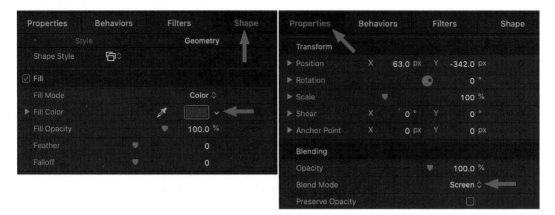

FIGURE 15.19 These two separate configuration screens in the Inspector modify the rectangle. The Shape text button changes based on what's selected in the Layers panel.

Play your project and notice how the animation in the background is reflected in the rectangle. The blend mode means they are sharing textures, which makes these elements look more organically related.

Let's add a gadget. These are animated whirly-gigs that do nothing except add movement and visual interest to an image.

1. Select the Shapes group.

2. Choose Library > Content > Gadgets.

3. Drag anything you like into the lower-left corner of the project. I chose Disk 02.

4. Disk 02 is added to the Layers panel, above the rectangle.

5. Press Shift and Option and drag a blue dot to scale the image small enough to fit inside the rectangle. (Shift constrains the aspect ratio, and Option scales from the center.)

6. Apply a Screen blend mode and watch what happens. Blend modes are used *constantly* in motion graphics, even if I don't specifically mention it for each example.

7. Play your project.

We could add more gadgets, drawings, or gauges—they all work the same way. In fact, take a few minutes to play with some different options. Almost all of them, especially drawings, contain animation, and they all look great with different blend modes applied (**FIGURE 15.20**).

I'll wait.

• • •

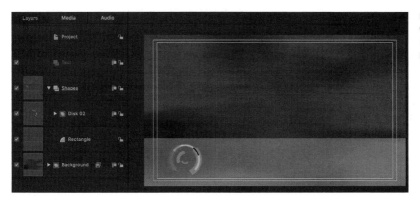

FIGURE 15.20 We are done with the bottom two layers. Here's what things look like so far.

Adding Text

Let's wrap this exercise by adding some text. Select the Text group then select the Text tool (Shortcut: T). Click in the top half and type **Model Railroad Expo**. Why? Because, uh, this chapter is about "training." Plus, I have some cool model train footage we'll add shortly.

All the text settings are defaults except these four:

- Set Font to Shabash Pro (or another casual font that you like).
- Set Size to 138.0.
- Set Alignment to Center.
- Set Line Spacing to −41.0.

Add Drop Shadows to Text

Drop shadows for text are important in Photoshop, but they are *essential* in video. Resolutions are lower, text appears for only a brief period of time, and backgrounds tend to be busy.

To add drop shadows to text:

- Select the text.
- Choose Inspector > Text > Appearance and scroll to the bottom.
- Check the box to enable drop shadows.
- Change Opacity to 95%.
- Set Blur between 3–5.
- Set Distance between 5–15.

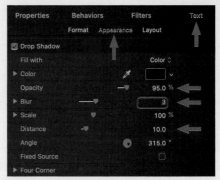

We can add drop shadows to *any* group or element in Motion, not just text. Element and group drop shadows are added and modified in Inspector > Properties.

Get in the habit of saving Motion projects frequently (Cmd+S).

FIGURE 15.21 All elements and formatting are complete. Now it's time to animate the text.

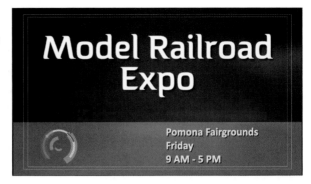

Again, with the Text group selected, add the Call to Action, which describes where the Viewer needs to go to see these wonderful trains. Format and color this text as you see fit—I used Calibri (**FIGURE 15.21**).

Adding Animation

Animation is the last step in this project, but it is also the most fun.

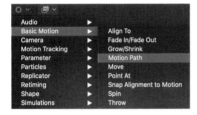

FIGURE 15.22 These Basic Motion behaviors are used constantly to get nontext elements to move.

Adding Behaviors

Select the "Model Railroad" text in the Layers panel. Click the Behaviors icon at the top and choose whatever text animation you like. (I used Text-Energetic > Dolly In.)

For Pomona, we'll do something different. Select the Pomona text layer. Choose Behaviors > Basic Motion > Motion Path (**FIGURE 15.22**).

FIGURE 15.23 Motion Path provides a starting point, an ending point, and a path along which the Pomona text moves.

The Motion Path behavior provides a path with a starting point and an ending point along which the selected object moves (**FIGURE 15.23**).

This path needs a bit of tweaking. Press the Home key to jump the playhead to the beginning of the project. Select the Motion Path effect; then, choose Inspector > Behaviors, and change Direction from Forward to Reverse. This means that the text will start outside the frame and slide in.

Drag the red dot located outside the frame so that the starting point of the text, indicated by a white box, is just outside the right edge of the frame.

Now, when you play the project, the text, uh, just *creeps* into the frame. This is way too slow. We need to fix this, which we will do shortly.

Adjusting Timing

Another important concept in Motion is that clips and effects are separate objects, each of which can have a unique start, end, and duration. By default, Motion places all effects and elements at the beginning of the Timeline. But in the case of this Call to Action, we want it to start later. That's where the Mini-Timeline comes in.

Move the playhead to the point where you want the text to enter. For this example, I want the text to enter when the timecode display at the bottom of the screen shows three seconds (3:00).

Select the Motion Path layer and press I. This trims the start of the clip so that it begins at the position of the playhead. This is called "setting an In," which **FIGURE 15.24** illustrates. Since we want this movement to last for only a second, move the playhead to timecode 4:00 and press O. This means that the movement, the effect, of the Pomona text will start at three seconds and finish at four seconds.

FIGURE 15.24 Press I to set an In (for the start) to a clip or effect. Press O to set the Out (for the end). Both marks are set based on the position of the playhead.

Move the playhead back to 3:00. Select the Pomona text layer and press I. This trims the text so it starts at the same time as the Motion Path effect. (Because the text and effect are different objects, you can have them start at different times, should you want to.)

To zoom into the Viewer, press Cmd+[plus].

To zoom out of the Viewer, press Cmd+[minus].

To fit the image into the Viewer, press Shift+Z.

To show the image at 100 percent size, press Option+Z.

These settings only affect the display in the Viewer, not the scaling of the clip itself.

We create something that attracts the eye then, a few seconds later, create something else that attracts the eye to continuously hold the viewer's attention during the video.

Now, when you play the project, the title animates on first, attracting the eye because it is big and bright and moving. Then, just as our eye starts to look for something else, the Pomona text slides on, grabbing the eye again and telling us where we need to go to watch model trains.

This is a good example of implementing the Six Priorities. We create something that attracts the eye then, a few seconds later, create something *else* that attracts the eye. By doing so multiple times, we can capture *and re-capture* the viewer's attention for the duration of our motion video.

Still, wouldn't it be great if we could add audio to drive the emotions then add video to make this motion graphic *really* exciting? Yup, you guessed it— we can, and it's next.

Interim Summary

Just to reflect what we've learned, this exercise covered:

- Creating and stacking groups
- Drawing and modifying shapes
- Adding and modifying gadgets
- Adding blend modes
- Adding and modifying text then holding it on-screen long enough to read
- Adding basic Motion behaviors
- Trimming clips and effects and adjusting their timing

Adding Media

Audio and video are central to almost all motion graphic videos. Let's continue with this project and add media to it.

Adding Audio

Audio is a key emotional driver in any video. Like any element outside the Library, we need to import the clip before we can use it. Let's continue working with the current project and add audio to it.

Choose File > Import (we *open* projects, but *import* media) and select the audio file you want to use. In the Mini-Timeline, a green audio clip appears.

When you import an audio clip, you'll see the green bar in the Mini-Timeline, but you won't see it in the Layers panel. That's because the Layers panel only displays clips and effects that affect the video.

We open projects, but import media.

Audio in Motion Is Extremely Limited

The audio tools in Motion are extremely limited. The best way to work with audio in Motion is to create a finished audio track before importing it. Trying to do an audio mix in Motion is an exercise in frustration.

The Audio panel controls volume and pan for an audio clip. To select a clip, click near its name in the Audio panel. Selected clips turn blue (**FIGURE 15.25**).

- To rename an audio clip, click directly on its name.

- To change the volume, drag the audio volume slider.

- To delete an audio clip, select it and press the Delete key.

- To change where the audio starts, drag the green bar in the Mini-Timeline to start at a new location.

FIGURE 15.25 To view or adjust an audio clip, click the Audio text button at the top of the Layers panel. To select an audio clip, click near, but not on, its name.

- To end an audio clip early, drag the end of the clip to the left in the Mini-Timeline.

To add an audio fade to the beginning or end of an audio clip, select the clip in the Audio panel; then choose Behavior > Audio > Audio Fade in/ Fade Out. Again, you won't see the audio effect in the Mini-Timeline, which just drives me *nuts!*

Start Audio 10 Frames Late

Because the web has a habit of chopping audio located at the very beginning of a clip, I've developed the habit of starting all audio ten frames *after* the start of a project. (This means the first ten frames are silent.) To shift an audio clip, select it in the Audio panel then drag the green bar in the Mini-Timeline to the right until the In starts ten frames after the beginning of the project, as shown.

If the audio is synced to video, trim the video to start ten frames early so you don't move the audio out of sync.

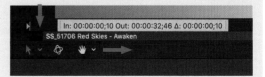

FIGURE 15.26 Click the HUD icon in the top-right corner of the Motion interface to display a floating, interactive control panel (Shortcut: F7).

The contents of the HUD change depending upon what is selected in the Layers panel. Select different elements—especially text—and watch how the HUD changes.

To disable an audio fade, drag the vertical line for the fade you want to turn off to a 0 frame duration.

FIGURE 15.27 The two vertical lines control fade duration (left for the opening fade, right for the ending fade). The numbers indicate the duration in frames.

To add fades to video, choose Behaviors > Basic Motion > Fade In/ Fade Out. The operation of the HUD for video is identical to creating audio fades.

Instead, select the audio clip, and click the HUD icon in the top-right corner of the Motion interface, as shown in FIGURE 15.26. (HUD stands for "heads-up display.") Or choose Window > Show HUD. This displays a floating, interactive controller which can be used for elements or effects.

Using the HUD, drag the vertical line on the left of the HUD to adjust the duration of a fade at the beginning of a clip; the line on the right adjusts the duration of the ending fade. If you don't see this fade control, click the faint two-headed arrow at the top right next to the title of the HUD, illustrated by the upward-pointing red arrow in FIGURE 15.27. This menu determines what is displayed in the HUD.

The HUD is a useful tool because it presents most common adjustments in an easy-to-use fashion. The Inspector offers more options, but the HUD is faster to use.

Again, just to stress, the best way to work with audio in Motion is to import a clip that is complete for duration, levels, and content—in other words, finished.

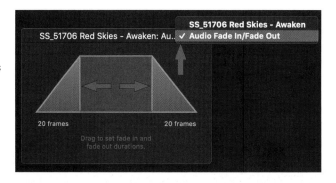

Adding Video

Working with video in Motion is more fun than audio because we can see what we are doing. To keep ourselves organized, create a new group—called "Video"—and place it at the top of the Layers panel. Select the Video group. Choose File > Import and choose the file you want to work with.

FIGURE 15.28 The video is imported, scaled smaller, and moved to one side. *(Train images courtesy of Model Railroad Builders—www. franandmileshale.com.)*

Video is always imported at 100 percent opacity and scale, which we rarely want to use in motion graphics. With the video clip selected in the Layers panel, choose Inspector > Properties and adjust the scale to decrease the size of the image.

Remember, images in the center get boring quickly, so, after scaling, I moved the video to the side. I also expanded the title to three lines using the Text tool to make room for the video (**FIGURE 15.28**). To continue holding the eye of the viewer, let's delay the start of the video until after the title animation finishes then have it quickly fade in.

To find the end of the title animation, select the text behavior in the Layers panel then put the playhead at the end of the Dolly In effect in the Mini-Timeline. Without moving the playhead, select the video clip in the Layers panel and press I. This sets the start of the video clip. Next, select the Fade In behavior in the Layers panel and press I again. This sets the start of the fade to match the video (**FIGURE 15.29**).

Play the project and watch the timing of your elements.

While there is no limit to the number of elements you can add to a Motion project, adding a lot may require rendering for smooth, full-speed playback.

FIGURE 15.29 Select the title animation in the Layers panel. Put the playhead at the end of the title animation ("Dolly In"). Select the video clip then press I to set the In.

For **FIGURE 15.30**, I decided to polish the video a bit more—I imported a second video and made it smaller. Then I added drop shadows (Inspector > Properties > Drop Shadow) to both video clips.

You may not be creating promo videos for model railroads, but the creation process is similar for whatever content you need to create.

FIGURE 15.30 Here's the finished video with a second video clip and drop shadows added.

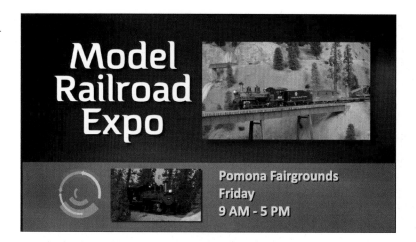

The Last Step: Saving and Exporting Your Work

Once you have your project created, you need to save the project then create a video; the two are not the same thing.

To save a project, which means you can open and modify it later, choose File > Save As (or File > Save, if you saved it at least once already). Remember, Motion does not save projects automatically.

To export a finished video, choose File > Share > Export Movie. When exporting ("sharing") a video, we are creating new media based upon all the elements in our Motion project. Motion master files are really big—often hundreds of megabytes. While you need to compress this for distribution, you always want to export a movie at the highest possible quality so you have a great-looking master file to use to create additional compressed files later as needed.

Using the Share settings, shown in **FIGURE 15.31**, you can:

- Export audio, video, or both.
- Determine the codec. My recommendation is Apple ProRes 4444 for a master file or H.264 for a compressed file for distribution.
- Color space. Leave at the default.
- Color channels. Leave at the default.
- Set Duration to the entire project.
- For Action, pick what works best for you. I generally set this to "Save only."

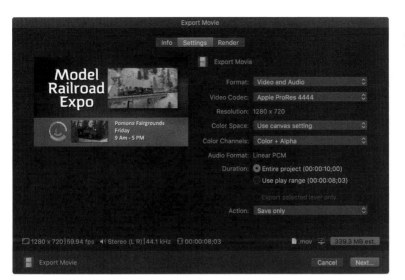

Export Movie

Info   Settings   Render

Export Movie

Model
Railroad
Expo   Format: Video and Audio

                   Video Codec: Apple ProRes 4444
    Pomona Fairgrounds
    Friday   Resolution: 1280 x 720
    9 Am - 5 PM
                   Color Space: Use canvas setting

                 Color Channels: Color + Alpha

                   Audio Format: Linear PCM

                      Duration: ● Entire project (00:00:10;00)

                                 ○ Use play range (00:00:08;03)

                                   Export selected layer only

                        Action: Save only

1280 x 720 | 59.94 fps   Stereo (L R) | 44.1 kHz   00:00:08;03   .mov   339.3 MB est.

Export Movie   Cancel   Next...

FIGURE 15.31 Share export settings used to create a finished movie.

Why Use ProRes 4444

ProRes 4444 is a 12-bit codec, supported in all major NLEs, that precisely matches the colors generated by a computer. Even more important for motion graphics, ProRes 4444 supports transparency in a clip, called the "alpha channel." This transparency information is retained when the clip is imported into any NLE.

After sharing is complete, and it may take a few minutes, double-click the movie file stored on your hard disk and enjoy watching your work.

3D Text

So far, all our text is flat—"2D," but Motion includes a very capable 3D text engine. **FIGURE 15.32** illustrates where we are going. Here's how to get there.

FIGURE 15.32 This is the finished, animated 3D text video we are about to create.

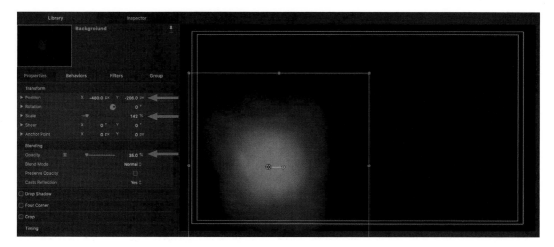

FIGURE 15.33 Place an animated background off-center then lower the opacity.

Create a new Broadcast HD 1080 project and two new groups. Label the top group "Text" and the bottom group "Background."

Select the Background group and drag Library > Particle Emitters > SciFi > Space Cloud into it. By dragging an element into the Viewer, you can position the effect anywhere, so aim for the lower-left corner, as shown in **FIGURE 15.33**. Select the Background group and choose Inspector > Properties. Lower Opacity to around 35 percent. I also increased the size to fill more of the corner.

Altering the opacity darkens and softens the cloud. (This will be a standalone movie with nothing under it, so I don't need to worry about the translucency that adjusting opacity creates.) You could put additional images or animated elements above it in the open space, but you already know how to do that. Let's concentrate on creating the 3D text.

Select the Text group on top; then click and hold the Text tool and select 3D Text. Click anywhere in the Viewer and type the **Order Now!** text. As with all on-screen text, keep the text short and punchy.

With the "Order Now!" text layer selected in the Layers panel, drag the text to the right side of the Viewer; then choose Inspector > Text > Format (**FIGURE 15.34**). Make the following changes:

- *Font*: Impact
- *Size*: 400 points (don't use the slider; type in the number)
- *Alignment*: Center
- *Line Spacing*: –100

The bounding box for particle systems and other animated effects changes over time. When scaling a particle system, play it a few times so you can see how much you need to change the size.

To my eye, 3D text looks best when using bold or black fonts. Think fat.

FIGURE 15.34 Prepping our 3D text with text format settings. (Safe zones are turned on.)

It's time to emphasize the 3D effect in the text. Select the text layer, choose Inspector > Properties, and click the triangle next to the Rotation parameter to display its settings. (I call this "twirling" a parameter.) Then, change Y to −45° and drag the position of the text in the Viewer to create a composition you like. Next, choose Inspector > Text > Appearance and make the following changes:

- *Depth*: 60
- *Front edge*: Concave
- *Lighting Style*: Backlit

Feel free to play with all these settings, illustrated in **FIGURE 15.35**, and watch how they change the look of your text. It is far easier to change a setting and watch what happens than for me to write pages of definitions. But, now, the *real* magic happens!

I like composing motion graphics with the Safe zone rectangles turned on mainly because they help me see where the frame ends and the dark interface begins. However, if these bother you, turn them off from the View menu in the top-right corner of the Viewer, or press an apostrophe (').

FIGURE 15.35 The Appearance settings determine the shape and lighting of 3D text. The next step is to add surface textures using the Material settings.

Change the Material popup menu from Single to Multiple. This applies different surfaces on the five edges of our text. Because this text is truly 3D, we can rotate it and see something different on all six sides (top, bottom, left, right, front, and back).

Given the choice, I like text with texture, provided it doesn't make the text too hard to read. All white all the time is boring. The Material options (**FIGURE 15.36**) in Motion provide lots of opportunity to play with the look of our text. (I especially like using concrete or stone surfaces applied to crumbling fonts like Chalkduster or Cracked.)

FIGURE 15.37 illustrates the settings I used for this example:

- *Front*: Miscellaneous > Motion
- *Front Edge*: Metal > Chrome
- *Side*: Plaster > Scraped Plaster
- *Back Edge*: Metal > Copper
- *Back*: Concrete > Aged Concrete

FIGURE 15.36 Click one of the Material buttons and choose from dozens of different surface textures in 11 categories. Experiment! See what you like.

At this point, you can apply any of the text behaviors we first learned about in Chapter 15. Fully animated 3D text! Because there are textures on all four sides, you'll see something different as you rotate the text 360°.

FIGURE 15.37 These are the Material settings I used for the text in Figure 15.32.

Animating 3D text is the same as animating 2D text. All the text animation behaviors work with both, as do keyframes. However, there is a category of behaviors available for all elements called Basic Motion that can also animate text. These include movements such as Fade in/Fade Out, Throw, Motion Path, and Spin that are useful for 3D text.

Because 3D text is truly 3D, as you change the Inspector > Properties > Rotation settings, watch how the lighting glints as the text rotates. Then, notice how the back of the text is different from the front. Just so you know, we can add custom lighting and cameras that zoom into and around each of these letters. But, that's for a different book.

As we explore the world of 3D, we need to understand the three axes, which use the mnemonic XYZ = RGB:

X Moves/rotates objects along the horizontal axis. (Color code: red)

Y Moves/rotates objects along the vertical axis. This tends to show perspective the best. (Color code: green)

Z Moves/rotates objects along an axis running perpendicular to the face of your computer monitor. This is the traditional axis around which objects rotate. (Color code: blue)

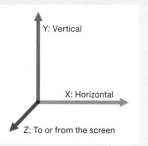

With the exception of 3D text and particles, Motion does not create 3D objects. All elements are displayed as 2D objects moving in 3D space. This is similar to moving a photograph around a room. The photo is moving in space, but when you look at the back of the photo, it does not display the backs of the people *in* the photo. This form of 3D is often called 2.5D.

KEY POINTS

Motion does so much and does it so differently that it is easy to feel overwhelmed. Motion is designed to create short, snappy video graphics. It is not a video or audio editor. Here are other key points from this chapter:

- Motion always works at the highest-possible quality.
- When creating projects for the web, the frame rate doesn't matter.
- Like Photoshop, the Layers panel controls placement of elements.
- Groups are used to organize elements and to allow a single effect to control multiple elements.
- The Inspector is where we make changes to any selected element.
- Behaviors are small, prebuilt animation modules.
- Filters are visual effects that change the look of an element.
- Text is styled in the Inspector and animated using Behaviors.
- The Mini-Timeline is where we adjust the timing of clips and effects.
- There's no limit to the number of elements in a project.
- Motion supports audio; however, the audio should be complete.
- Motion supports multiple video clips in a project.
- Keep all text short and punchy. Think haiku, not paragraphs!

Download the online chapter where we will look at specialized features in Motion that can add life and eye-catching visuals to your videos.

PRACTICE PERSUASION

Go back to the ideas you created for Chapter 1. If you have Motion, create a 15-second motion graphic based on your initial idea. If you don't have Motion, sketch a storyboard of your idea.

As you create this video (or storyboard), ask yourself:

- How did your idea evolve?
- How did you simplify your idea?
- What did you need to drop?
- What did you decide to emphasize?

When you have only 15 seconds, you can't do a whole lot. This means you need to intently *focus* on *one* message to *one* audience. Which did you pick—and why?

PERSUASION P-O-V

THOUGHTS ON LEARNING TECHNOLOGY

Larry Jordan

I still remember my first experience learning Motion. It was a training class, and it was awful. At the start of the first day, the instructor jumped right into 3D space. As someone new to motion graphics, it felt more like "alien space." I was lost and never did catch up.

I've used that experience in my teaching ever since. At heart, learning technology comes down to overcoming fear: the fear of not being smart enough to learn, which is nonsense. All of us are smart enough. The key is to get past the fear.

This is why I spend so much time covering the fundamentals. Once you understand how things work, you can more easily learn at your own pace.

This book covers a vast range of topics using some of the most sophisticated software on the planet. Give yourself permission to make mistakes, "creating garbage" I call it. Create something simply to learn how the software works, rather than to accomplish a specific purpose. I warn my students that the first thing they create will look awful. That takes the pressure off. Now they can have fun learning by creating trash.

Practice without a deadline, without anyone looking over your shoulder. Get oriented, discover how the software works, and play without trying to create anything worthwhile. As you do, opportunities will unfold before you. It's kind of like learning a new language; suddenly you discover there's a whole new community of people you can talk with.

The first goal in learning software is to understand its purpose. The second goal is to learn how it works. The ultimate goal is to use that software to help you persuade.

Persuasion is the goal—software is just a tool to get us there.

I sell sincerity and give
the truth away for free[1].

LARY LEWMAN
VOICE-OVER ARTIST AND "POETRYMAN"

CLOSING THOUGHTS

During the course of this book, we started with the big picture then looked at a wide variety of tools and techniques to deliver our message. We covered a lot. It is easy, especially given the last several chapters, to think that persuasion revolves around the tools we use to create our messages.

Easy, but wrong.

Our persuasiveness isn't determined by the software we use but the messages we create with it. These messages should be founded upon trust, delivered with enthusiasm, focused on a specific audience, and presented to grab the viewer's attention.

Our persuasiveness isn't determined by the software we use but the messages we create with it.

Here are seven thoughts on persuasion that flow through every chapter:

- Persuasion is a choice that we ask each viewer to make.
- Planning is essential. Design your story with a specific message for a specific audience with a clear and focused Call to Action telling them what to do.
- All messages are perceived as one-to-one, even if we are speaking to many people at once.
- Our audience is deeply distracted; our first goal is to capture their attention.
- The Six Priorities are core to capture and guide the eye of the viewer.
- Write less, and make every word count.
- The story and emotions in our visuals are as important as the words we use.

Even at our persuasive best, the viewer still needs to decide whether they will listen to us and follow our recommendations. We can't control their decision; we can only influence it. This is why the act of persuasion will always be a creative art.

[1] https://www.baltimoresun.com/obituaries/bs-md-ob-lary-lewman-20130715-story.html

BIBLIOGRAPHY

CHAPTER 1

Adams, Z. (2017). The Role of Thought Confidence in Persuasion. https://thedecisionlab.com/role-thought-confidence-persuasion/.

Ascher, S., & Pincus, E. (2019). *The Filmmaker's Handbook* (Fifth Edition). TheDecisionLab.com.

Booher, D. (2015). *What More Can I Say*. Prentice Hall Press.

Carnegie, D. (1961). *How to Win Friends and Influence People*. Simon & Schuster.

Charles, J. (2016). 27 Inspiring Quotes about Persuasion and Influence. http://www.artisanowlmedia.com/27-quotes-about-persuasion-and-influence/.

Eikenberry, K. (2014). Five Thoughts on Persuasion. https://blog.kevineikenberry.com/leadership-supervisory-skills/five-thoughts-on-persuasion/.

Gotter, A. (2019). 50 Call To Action Examples (and How to Write the Perfect One). https://adespresso.com/blog/call-to-action-examples/.

Nazar, J. (2013). The 21 Principles of Persuasion. https://www.forbes.com/sites/jasonnazar/2013/03/26/the-21-principles-of-persuasion/.

Norman, D. A. (2004). *Emotional Design*. Basic Books.

Overstreet, H. A. (1925). *Influencing Human Behavior*. W.W. Norton.

CHAPTER 2

Bang, M. (2016). *Picture This: How Pictures Work*. Chronicle Books.

Flowers, M. (2018). Does Sex Still Sell in the Age of Digital Marketing? https://www.ethoscopywriting.com/blog/does-sex-still-sell-in-the-age-of-digital-marketing.

Harrison, K. (2017). What is Visual Literacy? Visual Literacy Today. Retrieved Dec 28, 2019, from https://visualliteracytoday.org/what-is-visual-literacy/.

Lull, R. B., Bushman, Brad J. (2015). Do Sex and Violence Sell? A Meta-Analytic Review of the Effects of Sexual and Violent Media and Ad Content on Memory, Attitudes, and Buying Intentions. *Psychological Bulletin*, *141*(5).

Kay, Magda. Sex and marketing: how to use sex in your advertising, Psychology Today. http://psychologyformarketers.com/sex-and-marketing/

Marczyk, Jesse Ph.D., Understanding Sex in Advertising: Getting people to look or buy?, Psychology Today, Jun 26, 2017. https://www.psychologytoday.com/us/blog/pop-psych/201706/understanding-sex-in-advertising

CHAPTER 3

Booher, D. (2015). *What More Can I Say*. Prentice Hall Press.

Bridges, L., & Rickenbacker, W. F. (1992). *The Art of Persuasion*. National Review.

Clark, R. P. (2013). *How to Write Short: Word Craft for Fast Times*. Little, Brown and Company.

Dean, J., PhD. (2010). The Battle Between Thoughts and Emotions in Persuasion. https://www.spring.org.uk/2010/11/the-battle-between-thoughts-and-emotions-in-persuasion.php.

Embree, M. (2010). *The Author's Toolkit*. Skyhorse Publishing Inc.

Hausman, C., & Agency, D. L. (2017). *Present Like a Pro: The Modern Guide to Getting Your Point Across in Meetings, Speeches, and the Media*. Praeger Publishers Inc.

Khan-Panni, P. (2012). *The Financial Times Essential Guide to Making Business Presentations*. Financial Times/Prentice Hall.

Kilpatrick, J. J. (1984). The Writer's Art. Andrews, McMeel

Lewis, E. S. E. (1903). Catch-Line and Argument. Quoted in "What is the mysterious 'Rule of Three'? https://rule-of-three.co.uk/what-is-the-rule-of-three-copywriting/.

Newton, Isaac. (1704). *Opticks: Or, A Treatise of the Reflections, Refractions, Inflexions and Colours of Light. Also Two Treatises of the Species and Magnitude of Curvilinear Figures*, (Fig .12), Smith and Walford

Orwell, G. (1946). Politics and the English Language. *Horizon, 13* (Issue 76).

Rapp, Christof. (2010). *Aristotle's Rhetoric, The Stanford Encyclopedia of Philosophy*, https://plato.stanford.edu/entries/aristotle-rhetoric/#4.1

Sjodin, T. L. (2011). *Small Message, Big Impact*. Greenleaf Book Group.

Various (1977). *Reader's Digest Write Better, Speak Better*. Reader's Digest Association.

Weaver, K., Garcia, S. M., & Schwartz, N. (2012). *The Presenter's Paradox*. Journal of Consumer Research.

CHAPTER 4

Bringhurst, R. (2013). *The Elements of Typographic Style*. Hartley & Marks Publishers.

Garfield, S. (2011). *Just My Type: A Book About Fonts*. Gotham.

Loxley, S. (2004). *Type*. I. B. Tauris.

Selander, Kelsey. (1989) Bitstream Typeface Library marketing booklet.

CHAPTER 5

Budelmann, K., Kim, Y., & Wozniak, C. (2010). *Brand Identity Essentials*. Rockport Publishers.

Hurkman, A. V. (2011). *Color Correction Handbook*. Peachpit Press.

Lindstrom, M. (2008). *Buyology*. Broadway Business.

Norman, D. A. (2004). *Emotional Design*. Basic Books.

Pastoureau, M. (2001). *Blue: The History of a Color*. Princeton.

THESE BOOKS FURTHER EXPLAIN JEAN DETHEUX' IDEAS

Husserl, E. (1970). *The Crisis of European Sciences and Transcendental Phenomenology*. Northwestern University Press.

Merleau-Ponty, M. (2013). *Phenomenology of Perception*. Routledge.

Merleau-Ponty, M. (1964). *The Primacy of Perception: And Other Essays on Phenomenological Psychology, the Philosophy of Art, History and Politics*. Northwestern University Press.

Merleau-Ponty, M. (1992) *Studies in Phenomenology and Existential Philosophy*. Northwestern University Press. Particularly Cézanne's essay on "Doubt."

Hall, E. T. (1990). *The Hidden Dimension*. Anchor.

CHAPTER 6

Collins, J. (1998). *Self-development for Success: Perfect Presentations*. American Management Association.

Hausman, C., & Agency, D. L. (2017). *Present Like a Pro: The Modern Guide to Getting Your Point Across in Meetings, Speeches, and the Media*. Praeger Publishers Inc.

Khan-Panni, P. (2012). *The Financial Times Essential Guide to Making Business Presentations*. Financial Times/Prentice Hall.

Tufte, E. (1983). *The Visual Display of Quantitative Information*. Graphics Press.

CHAPTER 7

Norman, D. A. (2004). *Emotional Design*. Basic Books.

CHAPTER 12

Martin, R. B. (1986). *Stan Freberg: His Credits and Contributions to Advertising*. Texas Tech University.

Rose, J. (1999). *Producing Great Sound for Digital Video*. CMP Media, Inc.

CHAPTER 14

Hollyn, N. (2008). *The Lean Forward Moment*. New Riders Pub.

Murch, W. (2001*). In the Blink of an Eye*. Silman-James Press.

INDEX

NUMBERS

3D axes, in Motion, 373
3D bar chart, presentations, 134
3D text engine, Motion graphics, 369–372
48K vs. 44.1K audio compression, 279
180° Rule, 39–40

A

A (Action) in AIDA, 63
active words and phrases, using, 68
actors vs. talent, 150
additive color, 108
Adobe Photoshop. *See* Photoshop
advertisement, mission of, 64
AIDA (Attention, Interest, Desire, Action), 63
Aleck, Vicki, 236–237
analog-to-digital converters, audio, 260–261
angles, shooting from, 38
animated composite, creating, 354–352, 354–362
animating text, 352–354
animation, adding, 362–364
Animations panel, PowerPoint, 141
archiving projects, 341
Aristotle, elements of persuasion, 61
The Art of Getting Your Own Sweet Way, 62
aspect ratio, video, 228
attention, grabbing, 14, 16
audience
  addressing stories to, 16
  attention span, 65
  behavior, 10
  benefiting, 63
  creating desire in, 11
  defining, 11, 14–15, 59
  focusing on needs of, 56
audio
  adding effects, 270
  analog-to-digital converters, 260–261

bit-depth, 266
bus, 275
cables, 259–260
channels, 255, 265
codecs, 255
compression and output, 274, 278–280
controlling pan, 272
dialogue, 252
digital recording, 262–264
editing, 253, 265–269
effects, 273–278
enhancements, 335
exporting, 253
fades, 270
final mix, 338
frequencies, 255
frequency response, 254
headphones, 262
human hearing, 254
human speech, 254
levels, 254, 263, 270–273, 323
limiter, 336–337
measuring peaks, 267
meters, 267
microphones, 255–259
mixers, 261–262
mixing, 253, 270–278, 323
monitor speakers, 262
Motion graphics, 364–366
Multiband Compressor, 275–276
multichannel recorders, 261–262
music, 253
out-point, 266
output and compression, 278–280
Parametric EQ filter, 277–278
in-point, 266
pop filters, 265
power of, 253
record, 253
sample rate, 265
sound design, 253
sound effects, 253
sound waves, 253
test recording, 263

trimming, 266, 269
volume, 254
waveform, 254
workflow terms, 253

B

background
  blocking, 37
  presentations, 130–134
  image structure and effectiveness, 32
  and readability, 111
backgrounds and fonts, presentations, 130–134
backgrounds vs. layers, composite images, 195–196, 208
backlight, 151, 153–155
backups, video, 233
balance vs. symmetry, 44–45
Baltimore Children's Festival, 286
Bang, Molly, 31–33
bar charts, presentations, 136
behaviors, Motion graphics, 362–364
Benton, Michelle, 187
bigger, controlling viewer's eye, 29
bit-depth
  audio, 266
  color, 108
bitmaps, 106, 171–172
Bitstream, 74
black-and-white images, creating, 184
blackletter fonts, 85–86
blend modes, Photoshop, 209–211
blocking talent, 37–40, 158–159
blurring backgrounds, Photoshop, 206
Booher, Diana, 6–7, 11–12
Brand Identity Essentials, 103
brands and emotions, 12
brightness, controlling viewer's eye, 29
Bringhurst, Robert, 75, 78
bus, audio, 275
buying, influencing, 19–21